MARX TOYS

Robots, Space, Comic, Disney & TV Characters

Maxine A. Pinsky

With Values

Schiffer Publishing Ltd

77 Lower Valley Road, Atglen, PA 19310

DEDICATION

To MALCOLM KATES, M.D., Past National President of Antique Toy Collectors of America. Throughout the years, Malcolm has always shown interest and concern for my work, sharing detailed information on toys and offering his exceptional toy collection for photographs. As if he had not done enough, Malcolm introduced me to Mr. Peter Schiffer, the Publisher of this book.

and

To my dear husband, EDWARD, whose inspiring emotional and practical support and love helped immeasurably, not only with this book, but with my life.

Printed in China.
ISBN: 0-88740-936-9

Library of Congress Cataloging-in-Publication Data

Pinsky, Maxine A.
Marx toys: robots, space & comic characters: with values/ Maxine A. Pinsky.
p. cm.
Includes index.
ISBN 0-88740-936-9 (hardcover)
1. Toys--United States. 2. Louis Marx & Co. 3. Toys--Collectors and collecting--United States--Catalogs. 4. Toys--United States--Catalogs. I. Title.
TS2301.T7P494 1996
688.7'28--dc20 95-47997
CIP

Published by Schiffer Publishing, Ltd.
77 Lower Valley Road
Atglen, PA 19310
Please write for a free catalog.
This book may be purchased from the publisher.
Please include $2.95 postage.
Try your bookstore first.

We are interested in hearing from authors
with book ideas on related subjects.

Contents

Acknowledgments

Writing a book of this type is a collaborative effort and this book could not have been written properly without the considerable aid of many fine people. It is a pleasure to thank them for their help.

INTERVIEWEES

These former Marx Company employees most graciously shared fascinating and extensive input about the company and their unique experiences in it: Roy B. Ahlgren, Ed Furie, Raymond Hubiak and Seymour L. Wane. Their contributions are truly appreciated.

SPECIAL CONTRIBUTORS

The following generous individuals contributed detailed information, material, prices for toys, photographs and/ or allowed their valuable toys to be photographed. Their invaluable help is very gratefully acknowledged: David W. Allen, Jay Horowitz of MARX TOYS, Debby and Marty Krim of NEW ENGLAND AUCTION GALLERY, Ken Pfirman, Lloyd, Glenn and Jeffrey Ralston of LLOYD RALSTON GALLERIES, John S. Ritter, Herb and Barb Smith of SMITH HOUSE, Lenore Swoiskin, former Chief Archivist, and Victoria Cwiok of SEARS, ROEBUCK & COMPANY.

PHOTOGRAPH CONTRIBUTIONS

The following people very kindly allowed their toys to be photographed: The Allan Collection, Samuel Caley, Joseph Fanara, Rick Johnson, Malcolm Kates, M.D., Ralph Kaufman, Ed Owens, Trip Riley, Philip Rolin, Charles W. Weber, PhD. and Cheryl Weber, R.N.

PHOTOGRAPHERS

Special thanks to Robert and John Grubb for photographing many of these toys so beautifully.

PUBLISHER AND STAFF

Heartfelt thanks to Peter B. Schiffer and the staff of SCHIFFER PUBLISHING, LTD. for their part in bringing my book to fruition.

Last, but definitely not least, many thanks to my Editor, Jeffrey B. Snyder. As an author himself, Jeffrey was sensitive to a fellow author's feelings. Pleasant, cooperative and professional, he improved and clarified the text, placed material in it, and caught what would have been embarrassing mistakes. My feelings about his work can be summed up in one sentence: There should be more editors like him.

Finally, it only remains to say that I have endeavored to acknowledge everyone, but as the saying goes, "To err is human..." If anyone's name has been omitted, please accept my sincere apology. I am thankful for everyone's help.

Maxine A. Pinsky

Introduction

For years, the Louis Marx & Co. slogan for its toys was "Mechanical Toys That Are Durable And Mechanically Perfect." And, indeed, the author can attest to the truth of these words. Over time, she has (horrors!) accidentally dropped a few toys, which work just as well now as they did before falling. And many collectors can agree that Marx toys are amazingly long-running. Most perform almost as efficiently as when they were first purchased.

However, there is much more to Marx toys than the words of the Marx slogan. These glowingly colored toys are beautifully lithographed of tin, or finely modeled of plastic. The range of subject matter covered is staggering, from the realistic to the fantastic, but generally, Marx felt that children liked toys which reflected what was familiar to them. The company made almost every type of toy which performed almost every kind of action. Then, too, many Marx toys are designed to be humorous and it is difficult not to smile when watching them in action. In addition to all the above, Marx toys at the time they were produced were moderately priced. Unfortunately, this is no longer generally true, but some toys can still be found at fair prices. In considering the toys covered in this book, robots tend to be rather expensive, but some space toys can be bought reasonably. Then again, in character toys, the super-stars like Mickey Mouse tend to be on the high side, but lesser luminaries such as Pluto are cheaper. Ditto for metal toys as opposed to plastic toys.

In perusing this book, the reader might bear in mind certain information. Catalogs, particularly the Sears catalogs, have been used as research references. Needless to say, none of the toys mentioned in this book are currently available at Sears. While catalogs are very helpful, using them presents certain problems. Catalogs of the same year sometimes differ slightly in their advertisements from city to city. The difference could be in the price listed for a toy or even the presence or absence of a toy from one city to another.

China cup and saucer from Louis Marx's private service. Set given to author by Seymour L. Wane. Robert and John Grubb photo.

Other information to be aware of is the change in the Marx logos over the years. The first Marx logo from the 1920s consists of the letters "MAR" over a large "X" within a circle. The letter "A" is in the middle part of the "X." At the base of the circle, in small letters, is "Toys." On February 14, 1939, a new logo was published in the OFFICIAL GAZETTE. It resembles the earlier one, but has the additional words "Made In United States Of America" around the logo's circle. Another circle surrounds the words. This logo was used until 1959. At that time, Louis Marx finally decided to advertise more extensively. He went all out! In August 1959, a color advertisement extending from page 83 through page 94 was inserted in the magazine TOYS AND NOVELTIES. The advertisment was repeated in October and December 1959, too. This advertisement confidently promised "The Biggest, The Most Powerful TV Advertising Barrage The Toy Industry Has Ever Known." The Magic Marxie symbol was introduced, an appealing Pixie wearing a dapper cap. The Marxie has a head, arms and legs, but the rest of his body is in the shape of the new Marx logo. This logo consists of the circle and the words "Marx Toys." The corners of the large "X" are still there, but the "X" does not intersect. The Marxie proudly boasts that "As TV spokesman for Marx Toys, I'm going to sing my way into more homes, make friends with more kids and sell more toys than anyone in the history of the toy business!" "Wow!" (This advertisement is shown in the chapter entitled "The Ending of an Era.")

Another point for readers to consider is the increased use of plastic in later Marx toys. All the toys described in this volume are from the post-1950s, a time in which Marx intensified production of plastic toys. Tin toys were manufactured, too, in the post-1950s, mostly by Marx's Japanese subsidiary, Linemar. However, most of the tin toys have some plastic additions. For instance, a tin figure might have plastic arms and legs.

In the company's early use of plastic, toys were mainly tin with a few plastic parts, similar to the later Linemar toys. By 1948, toys with plastic parts were being advertised in catalogs and, by 1949, completely plastic toys began appearing. About 1952, full plastic production was started.

In conclusion, it is safe to say that as long as there are collectors, there will be interest in the marvelous Marx toys. Each generation rediscovers their charms.

Finally, it is the author's hope that readers will find this book enjoyable and useful.

The Ending of an Era

Louis Marx & Company at one time had been the largest toy company in the world. Marx himself became a millionnaire in his twenties. Featured in several prestigious magazines, he was a friend of President Eisenhower, top generals and many other famous celebrities. Marx's success is all the more fascinating considering that early in his career, he and his brother, David, shared an office so small that one of them had to leave when a visitor entered.[1]

Louis Marx was born on August 11, 1896 in Brooklyn, New York. In 1912, at the age of 16, he was already working for Ferdinand Strauss. The Strauss toys, ZIPPO THE CLIMBING MONKEY and the unfortunately named ALABAMA COON JIGGER, a clockwork minstrel, were two of the first lithographed tinplate mechanical toys mass-produced in America. In just four years, Marx was managing the company's New Jersey plant and he eventually became a company director.[2]

But Marx's place in the company was not to last. He and Strauss differed in establishing production policy. Marx felt that the company should emphasize production selling. Strauss did not agree and after a directors' meeting, Marx found himself "out on the street."[3] Marx then entered the Army in 1917 as a private. Naturally, he did not remain a private and by the time he left in 1918, he was a sergeant.

The outcome of World War I signaled the end of the competitive German toy manufacturers, who had severely limited the American toy market. Encouraged by the newly opened post-war toy market, an ambitious Marx, along with his brother David's help, turned his abilities to his own company. The two brothers became middlemen between the chain stores and the manufacturers. They would study how to cut costs on a toy, then get orders for the toys that they could produce at less cost. Since the Marx Company had no manufacturing facilities during this period, Marx paid other manufacturers to produce toys for them.[4]

In 1919, Louis Marx & Company incorporated and Marx took an office in New York City's Fifth Avenue Building, the small office mentioned previously. The company's central offices were to remain at this 200 Fifth Avenue location throughout Marx's ownership of the company.

By 1921, Marx acquired manufacturing facilities. When Marx's former employer, Ferdinand Strauss, sold his business, Marx purchased the dies for the two popular Strauss toys, the aforementioned ZIPPO THE CLIMBING MONKEY and ALABAMA COON JIGGER. Marx re-worked these toys, making them more colorful and bringing out larger models, resulting in sales of 8,000,000 units of each toy. Due largely to the success of these toys, Marx became a 26-year-old millionnaire.[5]

Marx's success continued when he observed a Filipino toy, to become known as the YO-YO. He sold more than a hundred million of them.[6]

In the mid-1920s, Marx achieved national fame. He acquired what were to become the three major manufacturing plants for the Marx Company by the 1930s. These were the Erie and Girard plants in Pennsylvania and the Glendale plant in West Virginia.[7] By the 1950s, Marx was licensing to factories in foreign countries.

Louis Marx is seen in the center of this photo shaking hands with the retiring L.C. Feist. March 17, 1957 photo from Erie Daily Times.

MARX FIRM OFFICIAL RETIRES

Officials of the Louis Marx and Co., 200 5th ave., New York City, the world's largest toy manufacturers, with plants including Erie and Girard, Pa., honored L. C. (Monty) Feist on his retirement as general manager, treasurer and assistant secretary of the firm in Pennsylvania. In the photo, Louis Marx, founder and head of the many Marx firms, bids Feist a fond farewell. William J. Keller (not shown) purchasing agent for Marx in Erie, was named to succeed Feist as manager of the Erie firm. Officials, from left, are Robert Lee, William Thompson, A. B. Marcus, Louis Marx, L. C. Feist, David Marx, George Dressler, Ed Hjelte and Joseph Pastore.

Louis Marx poses with Miss NBC. Playthings, April 1958, p. 83.

Photo of Louis Marx. Playthings, July 1952, p. 103.

[For additional information on Louis Marx's early years, see *GREENBERG'S GUIDE TO MARX TOYS*, VOLUME I, by Maxine A. Pinsky.]

Sadly, this success was not to last. Marx and his trusted associates grew older and Marx told a friend, "Toys are a young man's business."[8] Unfortunately, none of Marx's children were interested in taking over the company. Although Marx presumably spoke of his business to his family, still his unique way of running his plants did not help to foster interest. Marx ran the business from New York with only rare visits to his plants, one of the last being with his [second] wife.[9]

Some feel that although Marx made many fine plastic toys, his particular forte was tin toys. It has also been said that although Marx understood and accepted the need for safety regulations, he felt these regulations to be burdensome. In addition, although Marx finally did advertise in later years, he lost much by being slow to advertise on television.[10] As reported by *THE WALL STREET JOURNAL*, "Another problem was that with the exception of the BIG WHEEL, Marx has had few spectacular products in recent years. It missed the boat on space dolls and failed to capitalize on the electronic games craze. In the mid-1970s, a new SAFARI group of hunters, animals and jungle accessories just didn't catch on with youngsters."[11]

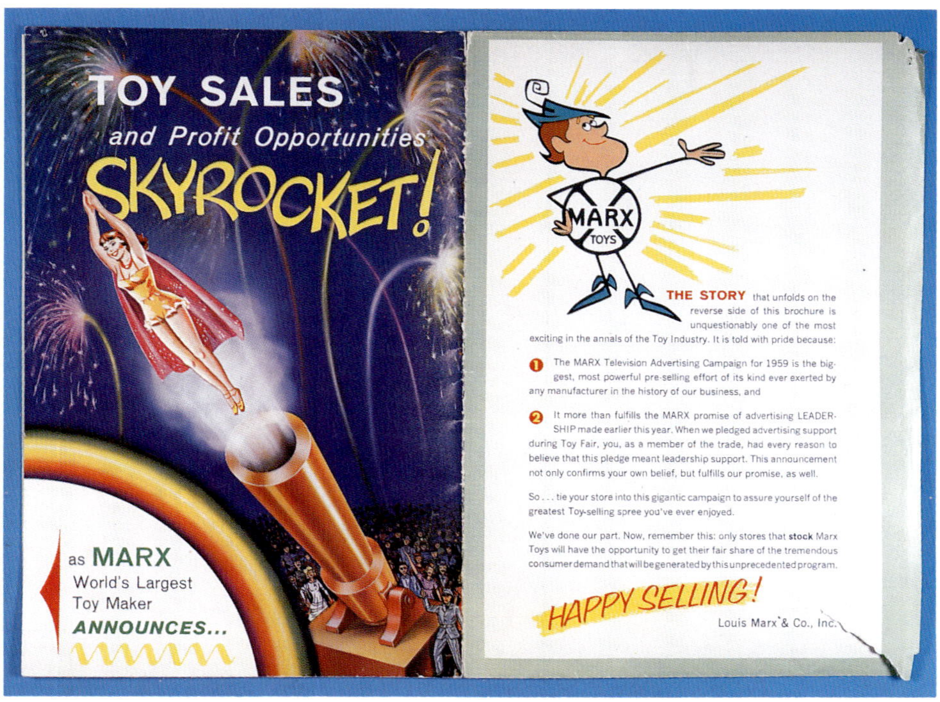

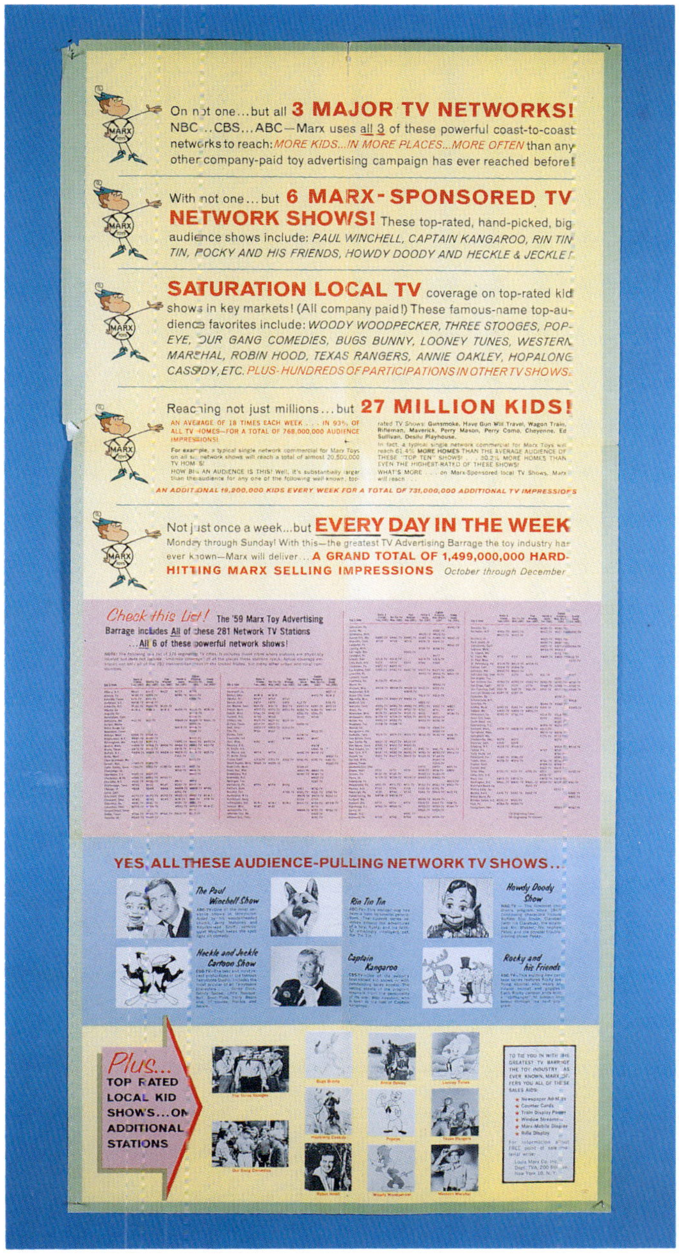

Part of an August 1959 advertisement (pages 83 through 94) from Toys and Novelties signaled Marx's overdue emergence into television advertising and introduced the Magic Marxie. Robert and John Grubb Photo.

In the 1960s, the company underwent the travail of at least two strikes, one in 1963 and another in 1967. The first and shorter strike lasted eight weeks and ended in the establishment of a labor-management committee to promote better relations, and increased pay and insurance benefits.[12]

The union struck again when the company would not accept the union's proposed new rate structure. This second and longer strike started in June of 1967 and not until April of 1968 was production scheduled to fully resume, although some operations had partially resumed.[13] All of these factors contributed to Marx wishing to divest himself of his company. In 1972, the deed was done, Marx selling his company to Quaker Oats for 51.3 million dollars. Unfortunately, the change in ownership was not a change for the better. As one article rather uncharitably put it, "Quaker Oats knows all about cereal but it has made a porridge of its diversification into toys. Since buying Louis Marx in 1972 for 51 million dollars (of which 14 million was for working capital) it has seen profits from toys shrink from a peak of 32 million dollars to only 3 1/2 million dollars."[14]

MARX GETS PELÉ: Soccer hero Pelé (R) recently signed a contract with Marx Toys. With him is Michael R. Meyers, vp research and development for Marx.

Marx vice president Michael R. Meyers signs soccer star Pelé. Playthings, November 1975, p. 100

Playthings

NEWS

The record of what's been going on with retailers, wholesalers and manufacturers

Quaker Oats names new president at Marx

CHICAGO—The Quaker Oats Company has put one of its toy division's top executives at the helm of the newly-acquired Louis Marx & Company.

Jack H. Asthalter, vice-president/marketing at Fisher-Price since 1964, becomes president of Marx, one of the nation's oldest and largest toymakers.

In another executive change, the Chicago-based manufacturer appointed Richard L. Harsch, Fisher-Price's vice-president/administration, executive vice-president at Marx.

Asthalter will report to Frank C. Schell, Jr., Quaker's group vice-president of toys and recreational products.

Both appointments, announced by Schell, were effective immediately.

Fisher-Price was Quaker's initial adventure into the toy field. It acquired the East Aurora, New York, company on September 1, 1969. Quaker reported sales of $678 million for the fiscal year ending June 30, 1971. For the nine months that concluded March 31, 1972, Quaker reported earnings of $27,278,000 on sales of $581,932,000.

Sales for the 1972 fiscal year are not yet available.

The parent company acquired

Richard Harsch is No. 2.

Asthalter heads Louis Marx.

Marx for approximately $51.2 million in cash. An agreement in principle between the two firms was announced May 10, 1972.

In a statement on the acquisition, Quaker said "approximately $37 million was paid for Marx's world-wide operating assets, including normal working capital requirements. Financing was obtained under a recently arranged, seven-year bank loan agreement."

Quaker paid an additional $14 million for seasonally expanded working capital.

The transaction is subject to the highly unlikely possibility of recision by Quaker, depending on results of an audit now under way.

Marx has long been known in the world toy market for its production of ride-em toys, games, trains, preschool, and other lines of toy merchandise.

For the past calendar year of 1971, Marx's world-wide sales approximated $67 million, with earnings of about $3 million.

The toy company's present senior management group, consisting of Louis Marx, David Marx, and Archie Marcus, will remain with the business in senior advisory capacities, Schell stated. All other members of the Marx management group are remaining.

Both Asthalter and Harsch will have offices in Marx's headquarters at 200 Fifth Avenue, New York City.

When Asthalter joined Fisher-Price in 1964, he brought to the famous preschool manufacturer a background of more than 30 years in the marketing of consumer goods. He has been responsible for all sales, marketing, advertising, and public relations.

He held executive posts at W. A. Sheaffer Pen Company, Wallace Silversmiths, and the Brunswick Corporation.

At Marx, the new president will have total responsibility for all aspects of the business.

Asthalter is married and has two sons and a daughter. He holds a B.S. in commerce majoring in marketing from the State University of Iowa, 1932.

The Asthalters presently reside in Orchard Park, New York.

Harsch, like Asthalter a native of Iowa, received a B.S. in business administration in accounting and an M.B.A. degree from the State University of New York at Buffalo. He is married and has three daughters.

Harsch joined Fisher-Price in 1959 and, since 1966, has been responsible for the firm's employee relations, information systems, profit planning and cost analysis. Prior to joining Fisher-Price, he had been controller with the State University of New York at Buffalo, where he also acted as a part-time faculty member in the evening division advancing to associate professor in management science and accounting.

Charles Weinschreider succeeds Asthalter as marketing director.

Ideal ends merger talk

NEW YORK—Ideal Toy Corporation and Artex Hobby Products of Lima, Ohio, have agreed to discontinue merger talks because they were not able to agree on terms. Last April, they had agreed in principle to Ideal's acquisition of the craft manufacturer.

Quaker Oats announces their new Marx president. Playthings, August 1972, p. 25.

Headquarters of the Quaker Oats Company at Stamford, Connecticut. Playthings, August 1973, p. 11.

Quaker Oats is certainly a worthy company and its toy company Fisher-Price has produced many good toys in its time. However, the Quaker Oats' way of running "Marx Toys" (as it was renamed), was markedly different from the way Louis Marx had run it. *THE WALL STREET JOURNAL* noted, "Instead it [Quaker Oats] tried to change Marx. The toy company's headquarters were moved from the Manhattan Building on lower Fifth Avenue, where many toy makers have offices, to Stamford, Connecticut, far from the toy buyers. Marx's toy-gun line was eliminated because it didn't fit the wholesome image Quaker wanted. And Quaker brought in executives from its other toy company, Fisher-Price."[15] Proliferation of executives was much of the problem. Former Marx employees have stated frequently that Louis Marx & Co. was a close-knit organization, almost like a family. As former employee Cal Cook stated, "Everyone knew everyone and they knew how the other person thought and what you were expected to do, almost automatically." Quaker Oats, which Mr. Cook considered a good company, destroyed the cohesiveness of the group. There were many changes of people in charge and they tried to make General Motors out of the Marx Company, which was impossible.[16] In 1975, the Erie plant was closed.

Marx advertisement in the form of a campaign button for the bicentennial election year. Marx was still a division of Quaker Oats Company at this point. Playthings, February 1976, p. 15.

In 1976, Quaker Oats gave up and sold Marx Toys to English-based Dunbee-Combex-Marx for 15 million dollars plus 3 million for working capital.[17] A plant letterhead from Quaker Oats stated, "Erie, Penn., April 15, 1976 - The Quaker Oats Company, Chicago, and Dunbee-Combex-Marx Ltd., London, announced jointly that D.C.M. has purchased the assets of Quaker's U.S. and Canadian Marx Toys business, as well as the Marx Hong Kong operation. The transaction in the U.S. includes the Marx Research and Development facilities, and Distribution facilities at Erie, Pa., as well as a plant at Girard, Pa., a plant at Glendale, West Virginia, and headquarters at Stamford, Conn. Jack H. Asthalter, formerly President of Marx Toys, has assumed the key new position of Vice Chairman, and Robert P. Butler of D.C.M. has assumed the role of President ... The company [D.C.M.] has existing operations in Australia, South Africa, Holland, Belgium, France and Germany, as well as a trading arrangement with the U.S.S.R."

Marx makes move to new showroom in N.Y.'s Merchandise Mart Building

STAMFORD, CONN. — Bob Butler, seated, president of Marx Toys, reviews architectural plans for Marx's new showroom with Larry Aledort, vice president of sales and marketing. The new room, which is under construction at the Merchandise Mart Building, 41 Madison Avenue, New York City, will occupy the 30th floor and was to be completed in November. The move was arranged to enable Marx to display its toys in a more modern and attractive surrounding.

New D.C.M. president Bob Butler moves the Marx showroom. Playthings, December 1976, p. 27.

A Quaker Oats memo of a press release, dated February 13, 1976, described D.C.M. as "Dunbee-Combex-Marx which is a diversified manufacturer of toys, toiletries and home-improvement products which has owned, produced and marketed Marx Toys in the United Kingdom since 1967; these were not part of Quaker's acquisition of Marx in 1972. Marx operations in Mexico in which Quaker Oats owns 56% interest are not involved in discussions." It is believed that Quaker Oats has since divested itself of the Mexican [called Plastimarx] plant.[18] The *ERIE DAILY TIMES* reported that, "D.C.M. was formerly an affiliate company of [Louis] Marx, but the toymaker sold to the London-based firm about 75% of Marx's European holdings ..."[19] Quaker Oats ended up taking a 23.2 million dollar pretax write-off.[20]

In her book, *AUTOMATA & MECHANICAL TOYS,* Mary Hillier gives a brief background of the new Marx owner: "Combex was the start of it all and may be said to have started with the aspirations of Max Reich, a refugee from Europe in the 1930s. Reich perfected a machine which molded combs from celluloid, saving the manufacturer the laborious job of cutting them from sheeting. He called his company Combex and prospered with the introduction of plastic toys. Meanwhile, following the Second World War, two young men, Richard Beecham and Basil Feldman, recognizing the emerging potential of the plastic materials for toy production, initiated the Dunbee Ltd. firm with a small capital and were able to take over both the company and services of Max Reich at Combex."[21]

NEW YORK — One of this season's most important visitors to the American Toy Fair was Richard Beecham, managing director of Dunbee-Combex-Marx, Limited. The London-headquartered firm is one of the largest manufacturers of toys in the world. He was in the United States to close the purchase of Quaker Oats' Marx Toys division. See page 17 for the news account.

A brief description accompanies the photo of D.C.M. managing director Richard Beecham. Playthings, April 1976.

D.C.M. had high hopes for their new acquisition because of their successful track record with Rovex interests which it turned from a loss into a 2 million pounds profit.[22]

For a while, all went well. In February 1978, D.C.M. acquired Aurora Products from Nabisco Food. D.C.M. felt that the Aurora toys, mostly the model auto racing sets, would blend well with Marx toys.[23] Aurora was moved to the Girard plant with resulting expansion. The Girard plant then experienced a growth in annual dollar value, employment went up and construction commenced. Former employees of the Erie plant, which was closed in 1975, were given hiring preference at Girard.[24] This promising start was not to last and D.C.M. began to lose money. *THE WALL STREET JOURNAL* reported that, "When Dunbee entered the U.S. Market in 1977, it had chalked up years of continuous profit growth. But net income slipped sharply in 1978 and in the 1979 first half, the company incurred a loss of more than 11.4 million dollars." Dunbee's businesses in Britain continued to be "successful as always."[25] Matters reached a point [in the U.S.] where the company slid into bankruptcy.

The reasons for the D.C.M. failure were varied. Some cited financial problems from disappointing Christmas sales and manufacturing problems in manufacturing Aurora racing toys.[26] Richard Beecham, the head of D.C.M., which by 1978 had become the world's third largest toy manufacturer, sent his bestselling doll, SINDY, to America, hoping to compete successfully against the BARBIE doll. BARBIE remained the winner while SINDY was the loser, helping the company farther along the way to ruin. A cartoon in *FORBES* even showed a salesman selling a SINDY doll with the comment, " ... and it cries like a Dunbee-Combex-Marx shareholder."[27]

Marx Toys' 1979 advertisements. Playthings, January-February, 1979.

Toys that work is one of the goals we take most seriously. A lot of good things are happening at Marx. Come on in and get the "Good News."

MARX®

"Nobody plays anymore. Everybody watches."

Involvement is the key to a successful toy, which is why Marx continues to be an industry leader. We make toys they get—and stay—involved with. We're introducing a number of involving new toys this spring. Come in and get the "Good News."

MARX®

"It's a dandy little game that teaches her that money isn't what it used to be for only $14.98."

In this age of consumerism, women's lib, and government regulations, people are finally thinking seriously about wholesome, well-made toys at reasonable prices. A lot of good things are happening at Marx. Come in and get the "Good News."

MARX®

See us at 41 Madison Avenue, New York, 30th floor

The *ECONOMIST* commented on another contributing factor to the company's problems. In referring to Dunbee's two acquisitions, Marx Toys and Aurora Products, the paper stated, "The two acquisitions were financed by loans. Dunbee-Combex also guaranteed 17 million dollars of Marx's debt and, at the end of 1978, total borrowings of 32 million pounds were almost one and a half times shareholders' funds, leaving little elbow room if things went wrong. Which they did. Production of Aurora toys, such as the electric racing car sets, were transferred to the two Marx factories in West Virginia and Pennsylvania, causing disruption just as the business was turning into the black. Then, last year [1979] profits in America were hit by high interest rates, falling demand for toys, and, at Louis Marx, stiff competition for its big selling lines like GREEN MACHINES and BIG WHEELS (both variations on children's bicycles)."[28]

As reported by the *ERIE MORNING NEWS*, Frank Chernek, former manager of engineering at the Girard plant, believed that while other factors could have led to the bankruptcy, the chief reason was the high cost of borrowing money. "We needed 25 million dollars to finance production. That's at the beginning of the year, and the receivables don't come in until the end of the year," Chernek said. "The short term loans financed production of the toys at the local plant. Marx sold these toys to stores, but didn't receive its money until the Christmas season. The short term interest rates were so high that the company was pushed into bankruptcy," Chernek believes. "We're talking of rates two or three points over the prime for our outside capital. When you're talking about 25 million dollars borrowed at these rates, it doesn't take long to do it," he noted. "If interest rates had not climbed dramatically since the Fall of 1979, the plant might not have been closed," Chernek added.[29]

Several laid-off employees offered additional reasons for the company's failure. One employee stated that materials were often less than the best and work in some departments was haphazard. Others believed the company management to be top heavy. It was felt that the latest toy line had also been hard to sell and overpriced.[30]

The Girard plant was closed on January 23, 1980, although a small maintenance crew remained for a while. Plans were made for the Marx facilities in Girard to be sold by a court-appointed real estate broker.[31] Eventually a number of businesses rented space in the Girard plant. The Glendale plant was also shut around this time. It was reported that a verbal agreement had been reached to sell the plant to Empire of Carolina.[32] The main Erie plant, which had been closed earlier, was bought by Erisco Wire Company. The Bucyrus-Erie Company, which manufactures construction and mining equipment, owns one of the other Erie buildings. (See Philip Zacks interview in *GREENBERG'S GUIDE TO MARX TOYS*, Volume I.) D.C.M. arranged for the sample toys stored at Erie, consisting of both Marx and non-Marx items, to be sold in a series of auctions circa 1977. (See Larry Passick interview, in Volume I also, for further information on these auctions.) As late as 1982, there were still auctions of molds, dies and remaining old toys of the plants, including the Glendale, West Virginia plant.

Finally, under presuure by creditors, D.C.M. consented in early February 1980 to operate under Chapter 11 of the Federal Bankruptcy Act. Chapter 11 permits a company to continue to operate under court protection from creditors, while it prepares an arrangement to settle with these creditors.

D.C.M. fared no better in England. After negotiations fell through to sell Marx and Aurora to the large American companies of Empire of Carolina, Inc. and Leisure Dynamics, D.C.M. went into receivership in February 1980. It was likely that shareholders would receive nothing.[33]

Empire of Carolina finally did buy all the Marx production equipment from the Glendale, West Virginia plant and the rights and tools for about 90% of the Ride 'Em products (including the BIG WHEEL). It is believed that Empire planned to use the name "Marx Toys Limited" (which it presumably purchased) on the Ride 'Em toys, but to date the Marx name has not been seen on the toys. The remaining tools were bought by American Plastic Equipment.[34]

In May 1980, the Marx Toy Company applied for a hearing in the U.S. Bankruptcy Court to sell 1.5 million dollars in assets to the Mego Corporation. Permission was asked to sell to Mego the SINDY and FIRST LOVE toy doll product lines, various game and playset inventories and molds, tooling and patents related to the aforementioned product lines.

The sale was one of a number of steps taken to liquidate assets. Estimates at this time of the company's debts to creditors and employees totalled about 100 million dollars.[35]

In June 1980, Aurora Products of Canade, Ltd. bought the Aurora Products line of toys for 5.65 million dollars. Assets bought were the Aurora products, patents, products made and in process, materials for production and special equipment. Toy product lines were also sold earlier to Amigo Toy Products.[36]

In March 1983, a new company reportedly purchased the name "Louis Marx Toys Int'l. Ltd." The company planned to have their products manufactured in Hong Kong.[37] A card from the new Marx company was printed with the Marx name and a Marx logo similar to the earliest logo with the letter "X" intersecting with the letters "MAR." Correspondence on April 7, 1983 from Herbert Sand, President of the company, stated that the Marx name was in the process of being acquired from Chemical Bank of New York.[38] This venture was not successful as will presently be shown.

And what remains of the Marx Company now? As it happens, the company is alive and well in Miami, Florida. Just how this turn of events came about requires a little background information. When Marx went into bankruptcy proceedings with Chemical Bank as the preferred creditor, all Marx assets were offered for liquidation. In September 1982, Jay Horowitz, President of American Plastic Equipment, Inc., Miami, Florida, purchased the assets of Marx Toys, which included only the tooling, not the rights. However, Mr. Horowitz's real interest was in the molds and to use the molds effectively, he needed the rights.

In 1983, Chemical Bank (still maintaining control over the Marx patents) tried to negotiate an arrangement with the Ting family in Hong Kong to begin manufacturing toys under the Marx name in the U.S.A. (This was the Louis Marx Toys Int'l. Ltd. company referred to above.) American Plastics initiated a lawsuit against Chemical Bank and the Ting family. Settlement of the lawsuit resulted in a three-way sharing of the rights among the Ting family, Chemical Bank and American Plastics.

It was not until 1988 that American Plastics was able to finalize control over the Marx rights, patents, copyrights and trademarks. The three-way deal among the Ting family, Chemical Bank and American Plastics had not worked out and the Ting family withdrew, enabling American Plastics to purchase the entire rights, molds, dies and remaining equipment of the Marx Company. Such an acquisition will allow Jay Horowitz to put into effect his plans for the future. He intends to reproduce certain toys from original Marx molds and to offer information from the Marx library. Some items have already been issued, including a Gold Rush Playset (never originally produced by

Marx as a "Playset," but the components were made from Marx molds), a plastic yellow taxicab, a metal Junior click pistol and blueprints for certain Marx trains. Jay Horowitz has offered assurances that the planned items will be made distinguishable from the original toys and that he will not allow confusion between the current issues and the older collectible toys.[39]

Actually, reissues (mostly in the form of plastic figures and some Playsets) are not new on the market. Briefly, some collectors like them because of their more reasonable prices, while other collectors fear that reissues may lower the value of their collections. Then, too, while experienced collectors can now recognize reissues by their color and possibly the quality of their plastic, in time these factors may become more like the originals and many collectors may be cheated. There is more that can be written both for and against the subject of reissues, but space limitations prohibit such a discussion.[40]

Because of reissues and since many of the older Marx toys are fortunately still available and sought after, interest in Marx will continue. Although the ending of the original Marx Company can be told in business and legal details, the Marx name will long continue to exist for all Marx enthusiasts.

FOOTNOTES

1. J.P. and Peggy McEvoy, "Talk With a Toy King" *READER'S DIGEST*, January 1955, p.126.
2. "The Little King" *TIME MAGAZINE*, December 12, 1955, p.96.
3. "Louis Marx: Toy King" *FORTUNE MAGAZINE*, January 1946, p.127.
4. "The Little King" p.96.
5. Ibid.
6. Ibid, p.94.
7. "Louis Marx: Toy King" p.163.
8. *WALL STREET JOURNAL*, February 8, 1980, p.6.
9. Author's interview with Edmund Galloway, October 27, 1983.
10. Information from former Marx employees.
11. *WALL STREET JOURNAL*, February 8, 1980, p.6.
12. *ERIE MORNING NEWS*, July 27, 1963, page unknown. Information submitted by Mr. K.
13. *ERIE MORNING NEWS*, March 26, 1968, page unknown. Information submitted by Mr. K.
14. *THE ECONOMIST*, April 24, 1976, p.108. Note: This is a British publication.
15. *WALL STREET JOURNAL*, February 8, 1980, p.6.
16. Author's interview with Cal Cook, October 29, 1983.
17. *THE ECONOMIST*, April 24, 1976, p.108.
18. Author's interview with Cal Cook, October 29, 1983.
19. *ERIE DAILY TIMES*, April 20, 1976, page unknown. Information submitted by Mr. K.
20. *WALL STREET JOURNAL*, February 8, 1980, p.6
21. *AUTOMATA & MECHANICAL TOYS* by Mary Hillier (London, England), Jupiter Books Limited, 1976, p.105.
22. *THE ECONOMIST*, April 24, 1976, p.108.
23. *ADVERTISING AGE*, June 5, 1978, p.8
24. *ERIE DAILY TIMES*, February 19, 1979, page unknown. The plant also manufactured pinball games and electronic games.
25. *WALL STREET JOURNAL*, February 20, 1980, p.21.
26. *ERIE DAILY TIMES*, February 5, 1980, page unknown.
27. "Tears in Toyland" *FORBES, INC.*, March 17, 1980, p.14. In this article, *FORBES* listed the two largest 1978 manufacturers of toys as General Mills and Mattel.
28. *THE ECONOMIST*, February 23, 1980, p.91.
29. Tony Pasquale, staff reporter, *ERIE MORNING NEWS,* May __, 1980, page unknown.
30. *COSMOPOLITE HERALD*, January 1, 1980, page unknown.
31. *ERIE MORNING NEWS*, June 10, 1980, page unknown.
32. *THE WHEELING INTELLIGENCER* as reported by Tony Pasquale in the *ERIE MORNING NEWS*, January 25, 1980, page unknown.
33. *THE ECONOMIST*, February 23, 1980, p.91.
34. Author's interview with Cal Cook, October 29, 1983. Telephone conversation with Empire of Carolina, June 25, 1985.
35. *ERIE DAILY TIMES*, May 2, 1980, page unknown. Article contributed to author by Mr. K.
36. *ERIE MORNING NEWS*, June 10, 1980, page unknown. Article contributed to author by Mr. K
37. *TOY AND HOBBY WORLD*, March 1983.
38. Correspondence to author from Herbert Sand, President of "Louis Marx Toys Int'l. Ltd." April 7, 1983.
39. The information about Jay Horowitz and American Plastic Equipment, Inc comes from two sources: John Torgerson, "The Jay Horowitz Story" *PLASTIC FIGURE & PLAYSET COLLECTOR*, vol. 8, August 1990, p.26. The second source is James R. Flynn, "Marx/Discovered in Miami/A Visit With Jay Horowitz of American Plastics" *CLASSIC TOY TRAINS*, December 1990, pps.82, 83, 85.
40. For a further discussion of the subject, see *the PLASTIC FIGURE & PLAYSET COLLECTOR*, starting with #7, June 1990.

The Marx Company Remembered: Experiences and Recollections of Marx Employees

PREFACE - INTRODUCTION

Facts and figures can give an accurate financial assessment of the Marx Company's fiscal situation, but only people can bring to life a picture of the company's history, of what it was really like to be there while history was being made.

The author was fortunate in being able to interview many former Marx employees, each extremely capable in his way. Included in this chapter is information provided by four of those interviewed. Although each employee encountered some difficulty on the job, as is true in any company, they felt their work to be stimulating and exciting. The length of time they chose to stay employed by Marx, varying from eleven years to twenty-nine years, would seem to bear out how they felt about the company.

If Louis Marx was the guiding light of the company, the brains behind it all, then employees such as the men included here were the company's backbone. They were the men who made it run and, as such, they contributed substantially to the Marx Company's greatness. The background of each individual is given and the men are listed in alphabetical order.

ROY AHLGREN

A copy of a photo showing the converted boiler room in which Roy Ahlgren worked. Mr. Ahlgren is seated in the back. Model maker Clark Rostatter stands guard with a toy rifle.

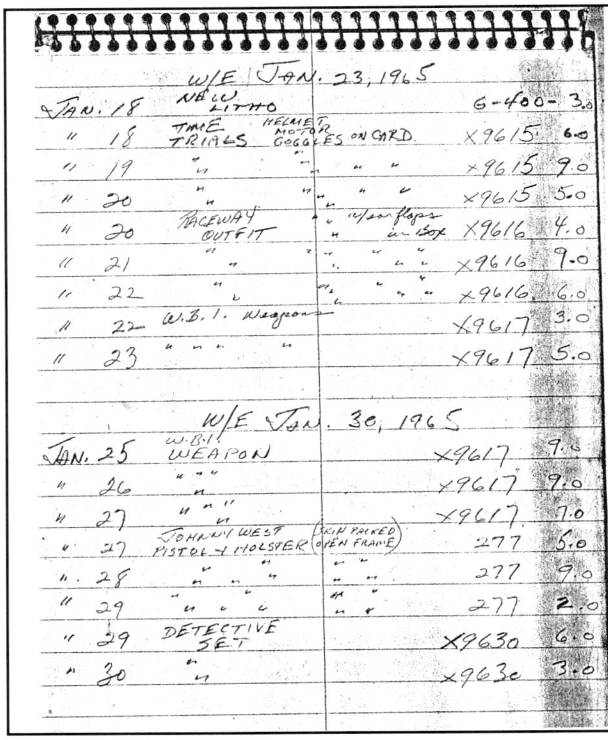

W/E JAN. 23, 1965
JAN. 18 NEW LITHO 6-400- 3.0
18 TIME TRIALS HELMET, MOTOR GOGGLES ON CARD X9615 6.0
19 X9615 9.0
20 X9615 5.0
20 RACEWAY OUTFIT w/ flags in box X9616 4.0
21 X9616 7.0
22 X9616 6.0
22 W.B.I. Weapons X9617 3.0
23 X9617 5.0

W/E JAN. 30, 1965
JAN. 25 W.B.I. WEAPON X9617 9.0
26 X9617 9.0
27 X9617 7.0
27 JOHNNY WEST PISTOL & HOLSTER (SKIN PACKED OPEN FRAME) 277 5.0
28 277 9.0
29 277 2.0
29 DETECTIVE SET X9630 6.0
30 X9630 3.0

One page of Roy Ahlgren's weekly records from the first five years of his employment.

The very talented Roy Ahlgren was born in Erie, Pennsylvania and studied art in local schools. He began employment with the Louis Marx Company in November of 1954. Art Director Jay Campbell looked over his portfolio and hired him to replace John Dahle, who had moved to California. Mr. Ahlgren worked eleven years at the Erie plant as an artist-illustrator. Afterwards, for seven years, he was an associate in a fine arts gallery. As an artist and patternmaker, he is nationally and internationally recognized for his efforts in serigraphy (silk-screen printing). He has a B.S. degree from the University of Pittsburgh and has taught commercial art and print-making for thirteen years.

Roy Ahlgren has produced over one hundred fifty extraordinarily beautiful limited editions of serigraphs over fourteen years. These works are in the permanent collections of numerous art centers, museums, colleges and private individuals. Mr. Ahlgren has exhibited in over one hundred juried shows nationally and internationally. His listings include WHO'S WHO IN AMERICAN ART, 1973, 1976, 1978 and 1980.

ED FURIE

Both witty and fascinating to talk with, Ed Furie candidly revealed the stimulating and sometimes bizarre world of the Marx Company. Mr. Furie started work at the company in September of 1949, as an office boy. He left in 1951 for the Korean War and returned in 1952. He then worked for Sales Manager George Dessler from whom he felt he received good training. "George Dessler was probably the greatest salesman in toys in those days. He taught everybody anything they wanted to know about how to sell toys." Mr. Furie then worked for David Marx and remained in the sales force almost twenty years. His jobs were many and varied. One of the most intriguing was tending to David Marx's Christmas list. "I was in charge of his Christmas list, taking care that all these big, important people in the United States got their toys. Yes, we spent a lot of money in those days. We used to have Paul Newman, Leo Durocher, Lorraine Day and Rockefeller on the list."

Mr. Furie started to work on this project around November, and it took him about two weeks. He would always obtain a list from Mr. Marx in case a name had been added to it, but nobody's name was ever taken off the list. "I'm telling you there were hundreds of names on that list, because it was every buyer in the country whom I took care of. It was not the idea to give them our toys, the idea was to go out and buy other toys. For his friend Sonny Werblin, who owned the New York Jets Football Team, he'd say to go out and get him some dolls for the football players. So, I'd go down to one of the local jobbers and ask how many Horseman Dolls did he have left? He'd say he had fifteen dozen ... this is prime at Christmas ... and I'd say I'll take them all. So, I'd send them out to the New York Jets Football Team." A hundred dollars would be spent on Lorraine Day, then a friend of Leo Durocher, manager of the Brooklyn Dodgers Baseball Team, but two hundred dollars would be spent on Durocher."

When he bought from a storekeeper, Mr. Furie might leave it up to him, but ask that the storekeeper not give him anything unwholesome. At other times, Mr. Furie picked out the items himself.

When Quaker Oats took over the company in 1972, Mr. Furie went with them and stayed for four years.

Looking back, Mr. Furie recalled that his first day on the job was almost his last. "I came home and told my mother that I can't believe this ... the man called me a dumb S.O.B. and I didn't pick up something and lay it over his head. My mother asked why didn't I quit? I said, 'No, I think I'm going to enjoy this place.' And I did. I stayed there twenty-six years."

RAYMOND HUBIAK

The affable Raymond Hubiak's background is interesting because it gives a good picture of what it was like at the Marx Company. Mr. Hubiak started work at the Girard, Pennsylvania plant in 1934. "Some of us fellows were hired right at the entrance door in those years by foremen needing help in their particular departments. This is the way I was hired." Mr. Hubiak's first job was on the night shift, snapping a gear guard on the Moon Mullins and Kayo Handcar. For this work he was paid $1.98 per night. He worked on various toys for several years and then became a line-helper on the line which assembled the sparkling mechanical locomotive with the "bell" sound. Mr. Hubiak's job consisted of bringing parts to the assemblies on the line. When World War II broke out, Mr. Hubiak was assigned to the stockroom and worked there for the duration.

After the war, Mr. Hubiak accepted a position in the Research & Development Department, a position in which he remained for the impressive length of twenty-nine years. He found the work in the R & D Department to be most challenging. "We designed and originated new toys, revised old items, built up models, created the parts lists for each item, made up sketches for the draftsmen and checked over the final blueprints.

"The most nerveracking part of all was meeting tight deadlines, since every project involved the above steps, especially with the items which were accepted for production."

Raymond Hubiak was responsible for many toys. He designed and built up the "Operating Box Car" in O gauge, and an "A Overhead Block Signal" in HO gauge. He also developed and built up "an electric and battery operated horn" for steam and Diesel locomotives, one for the steam sound and one for the Diesel sound. In addition, he worked on many revisions of existing items.

The name of WALTER HUBIAK, Raymond's brother, was encountered by the author several times in her research. Walter started work at the Erie, Pennsylvania plant in the late 1940s or early 1950s, eventually also winding up in the Research & Development Department. He was always interested in creative work, such as designing models and wood carving. Raymond remembers Walter working on a "Boxing Toy," "Penny the Poodle" and the "Stutz Bearcat Car" plus many other items. Besides working all day at the plant, Walter would assemble many toys at home which were packaged in kits for store samples and salesmen. Walter loved his work but health problems forced him to terminate his employment at Marx. He passed away in 1972.

SEYMOUR L. WANE

The dynamic Seymour L. Wane started working for Louis Marx in 1948 as an accountant. He was brought into the company at the request of Archie B. Marcus, Marx's Vice-President. In 1968, Mr. Wane himself became a Vice-President, the only person to rise so high, except for the founders, Louis and David Marx, and Archie B. Marcus. As the financial and administrative Vice-President, Mr. Wane ran the office and was finally able to visit the factories. He worked "everywhere." Mr. Wane remained with the company until it was taken over by Quaker Oats, but was brought back to assist in negotiations with Dunbee-Combex-Marx because, "I knew where everything was hidden."

Mr. Wane felt that the most challenging part of his job was working with Louis Marx. "You never knew what he was going to ask you ... he would call you at 8 o'clock in the morning after he'd been up all night with his memos and studying, and he would fire questions at you." However, Mr. Wane believed that Louis Marx was very loyal to his employees and this loyalty was returned to him. Very few people ever left the company.

Mr. Wane felt that the work was very stimulating. "Every year was a new challenge. We brought out a new line of toys at the start of every year; we started from scratch which was very exciting. And the people, the people were terrific, too."

EXPERIENCES AND RECOLLECTIONS OF EMPLOYEES

The core of any toy company is its toy line and Marx Toys are among the best and most popular toys ever made. No one is more important in contributing to the quality of a toy than the artist, but just what does an artist do? How much control does an artist have over his or her work? Artist-illustrator Roy Ahlgren provides the answers to these questions and much more information as well.

THE "ART" ROOM

As Roy Ahlgren describes it, the "Art" room was located in a converted boiler room approximately twelve feet by thirty feet. Five by five foot stalls had been built down either side of this art room, two on the West side and three on the East side. The entrance was just ahead of these three stalls, which were about forty inches high. The front (West) end of the room was occupied by Jay Campbell, facing the others. Jay was in charge of the artists.

JOB TITLES

Mr. Ahlgren explained that, "As all things evolved in this company, no one really seemed to have official titles. People just moved themselves into these positions throughout the company. When I was hired, they seemed to vacillate and finally decided that my title should be 'artist.' Prior to this time, the artists were thought of as model painters. All the emphasis was on the model room, although development took place, that is, concepts came from many sources. Jay Campbell was a very creative man who introduced many good 'sell' concepts (presenting the best visual appearance through copy and pictorial usage) as well as designs.

"Ray Lohr, the Research & Development Manager, worked with Louis Marx and C. Edward Hjelte, Director of Marketing, in producing the models. [For more information on Mr. Hjelte, see Preface and Chapter 4 of GREENBERG'S GUIDE TO MARX TOYS, Volume I.] Ray had almost thirty to forty model makers working with him. They would produce the mechanisms to create working models. The art room would create the color and packaging."

ROY AHLGREN'S WORK AS AN ARTIST

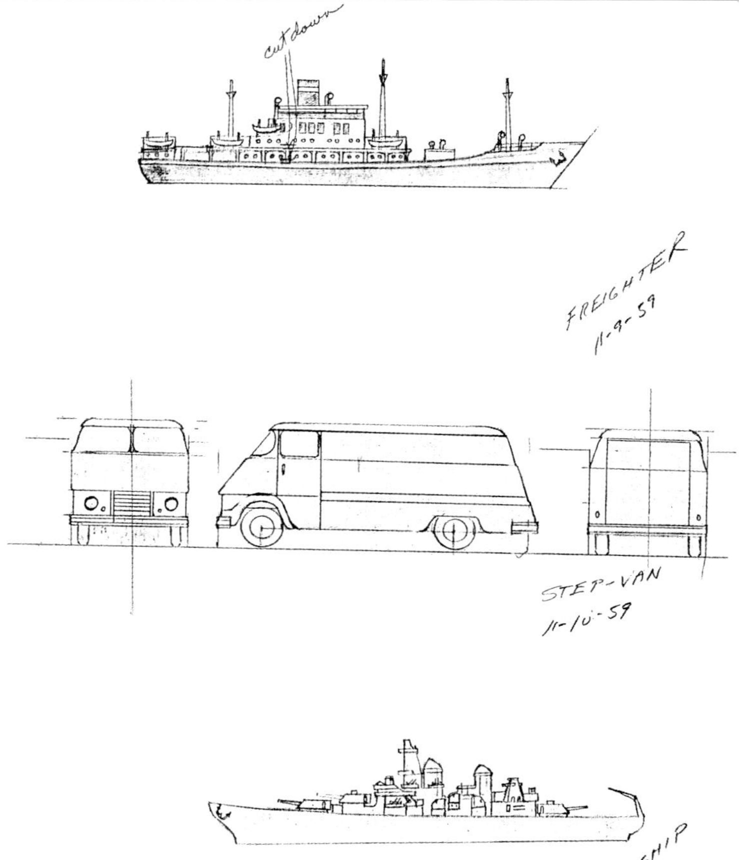

Artist Roy Ahlgren's first assignment, a series of ships and vehicles.

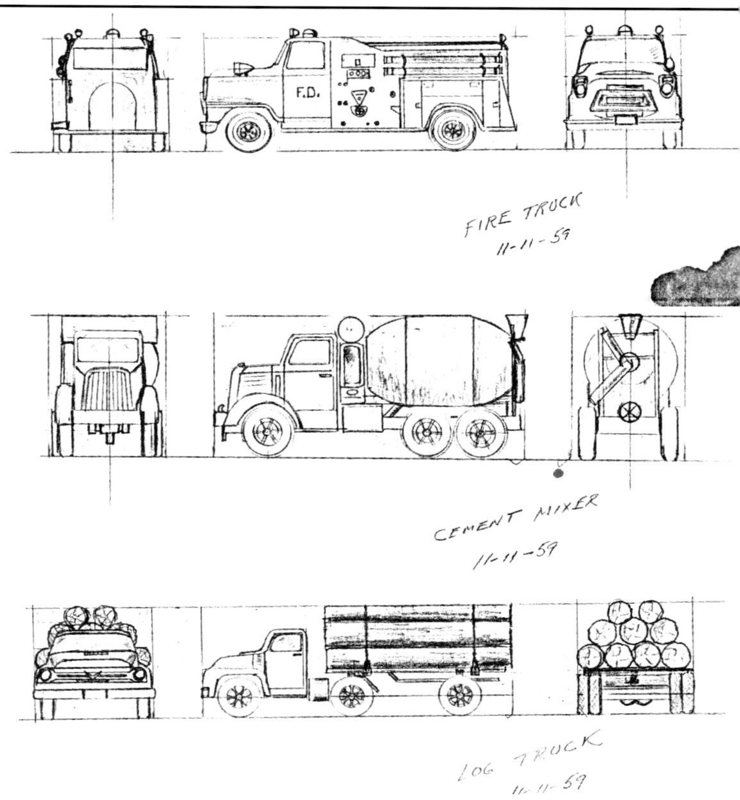

Trucks from Roy Ahlgren's first assignment.

"My job consisted of designing new concepts in toys, as well as working from the Art Director's suggestions. I also designed packaging and art, self-adhesive labels, 'Bagatelles,' and Arcades and Pinball Games. Due to the quantity of production and the size of the company, we could never be sure whether the items were produced, but much of my work was printed and produced.

"The Art Department did layout and prepared camera-ready art for instructional sheets, books and booklets, and frequently wrote the copy. At times I would be required to build special demos or point-of-sale displays for certain items, such as a truck or a cannon. There were always many cards to be designed and painted, to which guns, etc., were attached with rubber bands, or blister packaging. Many times I had to airbrush certain toys or hand-paint samples for the New York showroom. This meant that we had to paint a dozen or so card and box samples for the New York showroom where buyers would decide and put in orders which determined the runs on the assembly lines in the Erie plant.

"Developing new colors was always interesting, as was working in 3-D with clay when building a variety of items such as a doll head, a moon base and the seated body of 'Garloo' the 12 inch battery-powered walking monster."

LIMITATIONS AND CREATIVE FREEDOM

Sweepstakes label designed by Roy Ahlgren shows the vibrant range of color used by the Marx Company. Robert and John Grubb Photograph.

Usually, Jay Campbell would discuss the format and concept, sometimes suggesting the limitations of color and size as pertaining to cost. "Much of the time," Roy Ahlgren stated, "I was able and encouraged to create the art, lettering and color for box or card art. Some companies, such as Sears, wanted a certain corporate look to the box art and other items.

"On box art, a rendering of the toy with appropriate lettering and color to fit was my choice. In any event, the final decision was made in New York, although they usually had faith in, and accepted, the art we presented.

"When working with Disney, we were free to develop toys and packaging with few restrictions. My most memorable restriction came from New York when I designed a Mickey Mouse/Pluto Walking Toy which was rejected as pornographic." This toy, however, was produced after all. (For further description of toy, see "Ramp Walkers" in Disney chapter.)

Disney character walking toys designed and described by Roy Ahlgren.

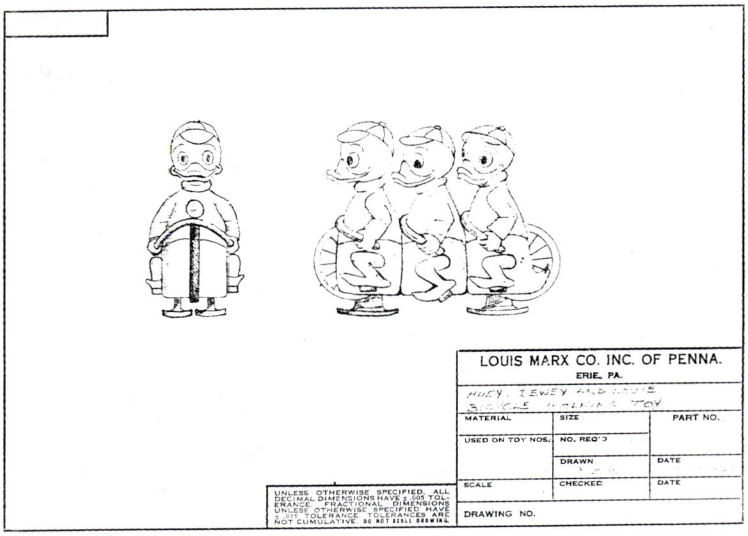

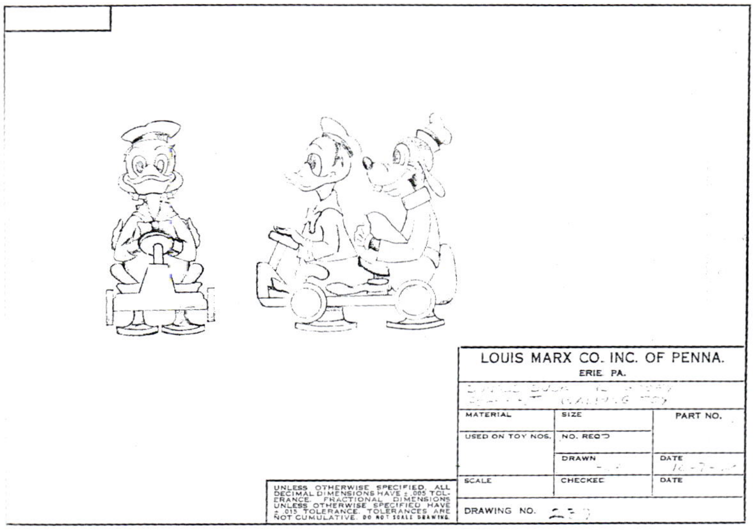

More Disney character walking toys.

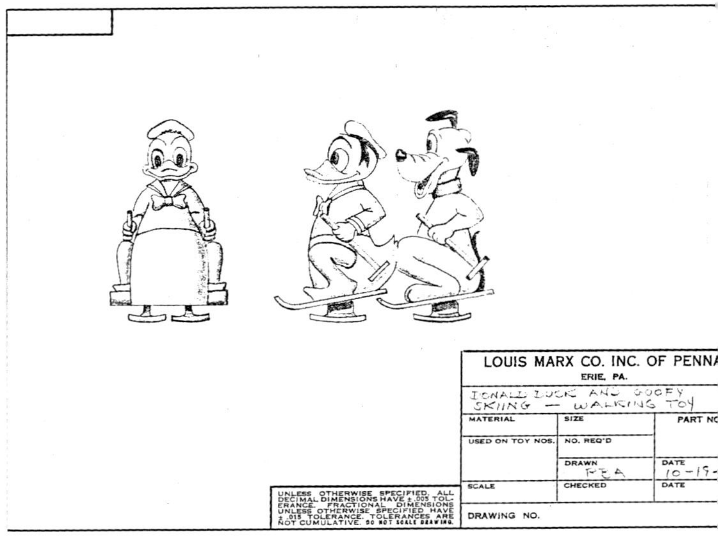

Fred Flintstone and Barney Rubble walking toys designed and described by Roy Ahlgren.

This Roy Ahlgren Pluto and Mickey Mouse walking toy was considered at first too risqué to produce!

"With television toys, we always attempted to represent and adhere to the original TV concept in order to have recognition in the marketplace."

DEVELOPING COLORS

Mr. Ahlgren was responsible for the art, lettering, coloring and the overall design of boxes, item cards and other art with final approval of the Art Director. "Mixing color was of prime interest to me and I developed a series of colors mixed in baby food jars with the lid tops painted to identify the color inside. My favorite colors were desert tan and some khakis for war toys which were soon to be in disfavor due to the Viet Nam War. This color mixing carried over into my fine art serigraph prints, which I continue to publish today and exhibit worldwide."

IDEAS FOR TOYS

Many toys were developed in the Model Room and sent to the Art Room for painting and packaging. Sometimes the item started from the artist's design. It was difficult to predict which source would produce results. Mr. Ahlgren stated that, "On coffee break one day, Carl Merl, a model-maker, and some of us observed a Bucyrus-Erie large crane across the street from the Marx plant. Carl then began to build the SUPER CRANE for which I did the box design. Other times, ideas came from competitors' toys or any other source. Secrecy was of prime concern in toy and art development."

The billboard was erected by the Keystone Box Company in Pittsburgh, Pennsylvania using Roy Ahlgren's box design. A design idea was given to Mr. Ahlgren by art director Jay Campbell in the other half of this picture.

BOXES

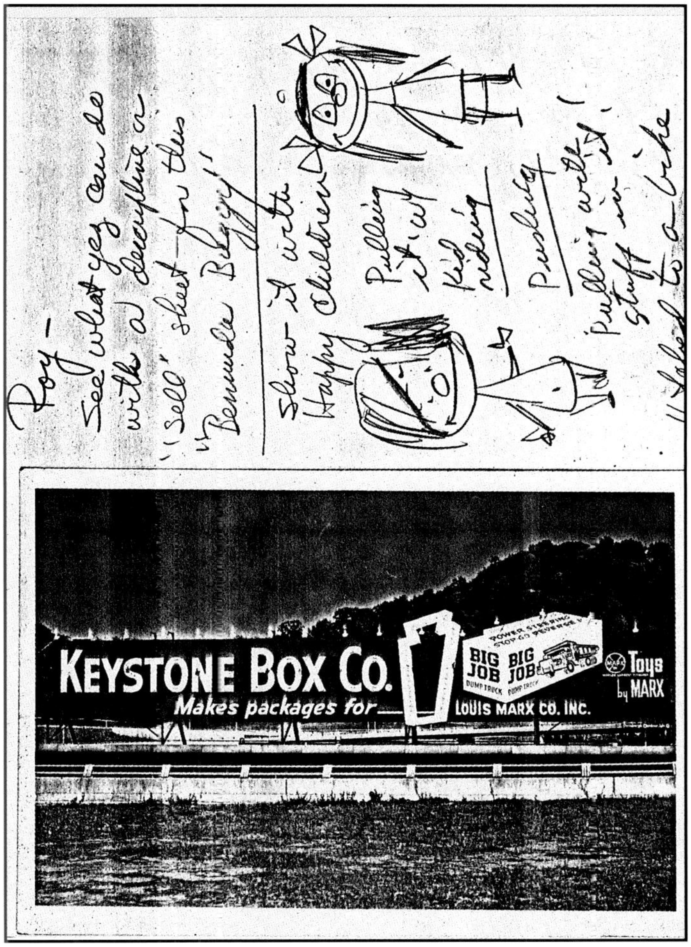

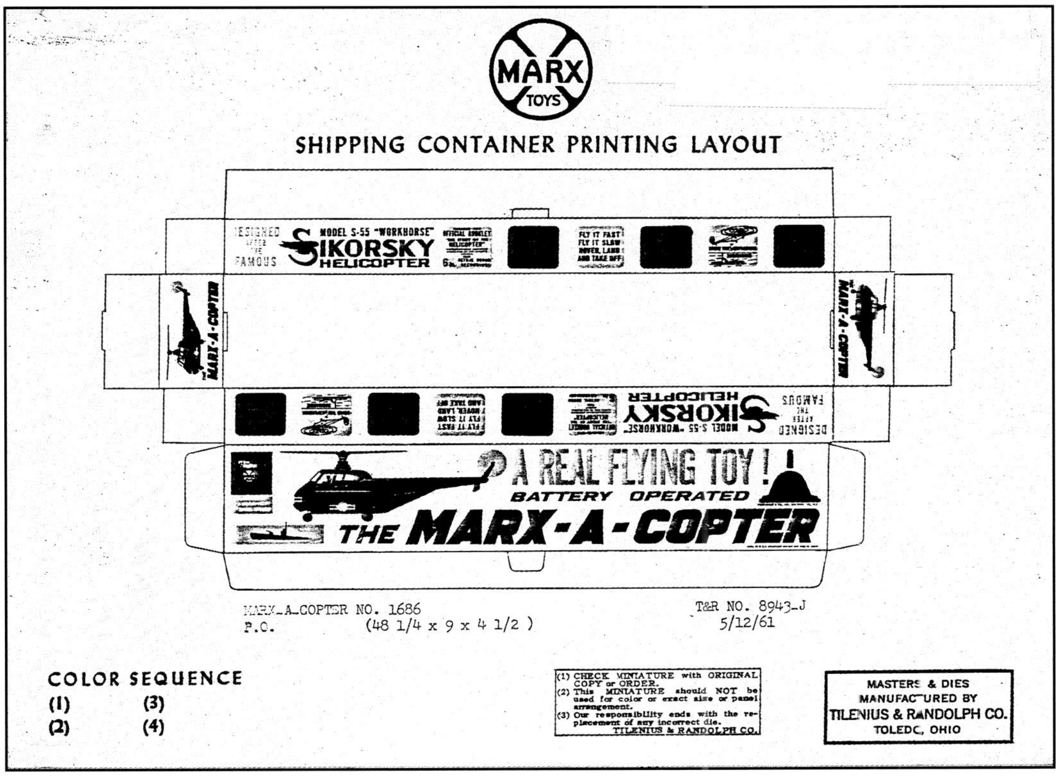

An example of Marx box art designed by Roy Ahlgren.

Toy enthusiasts have long noticed that box illustrations sometimes differ in small details from the actual toys. "I suppose we would enhance or exaggerate somewhat," observed Mr. Ahlgren, "but I believe this was the result of artistic style as opposed to photographically produced box art. We attempted to produce eye-catching box art to sell the toy. During this time (the late 1960s), the use of photos on the boxes in the form of large paste-on labels was coming into prominence. Again, the concern was with what the competition was producing."

The Art Department sometimes created the art on boxes for toys from foreign plants. Roy Ahlgren pointed out that, "Much of the art on toys from Hong Kong, England, etc., was already developed when it came to us, however there were other times when we would receive items from foreign plants where we did create the sell-art and packaging."

PLASTIC OR METAL

Mr. Ahlgren felt that plastic toys, as opposed to metal toys, allowed for a greater flexibility in development, although the charm of the lithographed metal toys was sacrificed. For instance, blow-molding opened a new approach to toy development. "Many times I was required to model items in three-dimension using plasticene clay. Some memorable toys were the blow-molded CROW with the fence-target game, the GARLOO battery-powered monster, the dollhead for TWINKIE the walking doll and the MOON BASE."

Roy Ahlgren's work was not without its problems, some of which were creative. "Jay Campbell handed a dozen small metal fish to me with a request for twelve different designs. I started a few and seemed to draw a blank." Nevertheless, the toy was produced. Another problem was time. Mr. Ahlgren was not overly happy with the fifty-hour week he was required to work, "However, our freedom to create made up for that."

FAVORITE TOYS

Mr. Ahlgren had many favorite toys for which he created the box art, including the Big Job Truck, Stutz Bearcat, Mystery Space Ship (see Space Toys chapter for photo), Big Loo, Electric Pinball and Hi-Light Scoop & Dump Truck. The Daniel Boone booklet was enjoyable to him as was painting the art for the Apollo Moon Bagatelle and others. "There were hundreds of exciting and enjoyable items of art in my eleven years with Marx Toys that made going to work a real pleasure for me."[1]

LINEMAR - MARX'S JAPANESE SUBSIDIARY

The Marx artists were always of importance to the company. Another group that came to be significant in later years, especially in the 1950s and 1960s, was Linemar, Marx's Japanese subsidiary. Linemar was actually a section of Marx-Japan which was a subsidiary of the Marx Company. Seymour L. Wane, former Vice-President of the Marx Company, explains Linemar as, "Items made specifically for us, very often using our dies and molds. Linemar was used for items made by others [the Japanese] that we imported."[2] Basically, Linemar consisted of a group of Japanese manufacturers who contracted with Marx to produce toys for the company under the Linemar label. Ed Furie, David Marx's assistant and part of the sales force, put to rest the rumor that Marx owned a factory in Japan when he stated that no American could do so. They [the Japanese] would have contracts. They were really contracting and they would manufacture our items for us and they probably manufactured their own items, also." Most of the time this worked out well, but once in a while there were problems. Mr. Furie recalls the time he went into a Woolworth's store and saw some sparking airplanes. He tested some and found that contrary to the contract, the airplanes did not spark. "He [the Japanese manufacturer] had shipped them without the sparks, or at least some of them without the sparks, and he made that extra couple of pennies on each, which amounted to a lot of money. Well, Louie told the Japanese [manufacturer] that he wasn't going to pay for them; I remember that very distinctly. The Japanese manufacturer said that wasn't a problem, it was a mistake, we'll manufacture your planes with the sparks. In the meantime, the manufacturer shipped all his aircraft first without the sparks at a much lower cost. Then when we were ready to ship ours, we had no one to ship to because he had taken all our customers."

As for the amount of control Marx had over Linemar in addition to shipping over the dies and molds, Mr. Wane states that Marx did not run the Japanese plant, but Marx agents would visit it and Marx could control the quality. Mr. Furie also comments that the Japanese could sell their own items under the Linemar name but "Marx would tell them what he wanted, too."

The effect of the relationship between Marx and Linemar resulted in some superb, if expensive, toys that are the delight of collectors. Many Linemar toys are detailed in this book.

THE MARX COMPANY AND FOREIGN LOCATIONS

Marx may not have owned any plants in Japan, but did own one in Hong Kong which, according to Mr. Wane, "... turned out to be extremely valuable." Rising costs in Japan necessitated the move to Hong Kong around the 1950s - 1960s. Unlike the case in Japan, some Marx personnel worked in the Hong Kong plant, including an accountant who still works for Mr. Wane. Marx Hong Kong toys were sold as far away as Canada.

Then there was the tale of the Taiwan plant. As Mr. Furie tells it, Marx did have a plant in Taiwan in the 1960s. Unfortunately, Marx and the Taiwanese people did not get along. Tempers escalated to such a degree that, finally, the plant was burned down and that was the end of the Taiwan plant.

The Marx Company is reported to have had factories in France, Brazil and South Africa. Mr. Wane explained that basically there were offices, not plants, in France and Brazil and just a representative in South Africa. The reason for the offices is that Marx toys were sold in these countries. Marx did own 100% of a Canadian assembly plant. Then there was the Marx plant in Swansea, Wales and, closer to home, Marx's Mexican subsidiary Plastimarx. The Marx Company had offices in Germany where items were made for them, such as dies and molds, metal toys and painted figures. Hong Kong later replaced Germany as a source for painted figures.

MARX DOMESTIC PLANTS AND LOCATIONS

The Marx Company's United States plants consisted of the Erie and Girard plants in Pennsylvania,and the Glendale plant in West Virginia. The Erie plant manufactured mostly mechanical wind-up toys, the Girard plant was generally known for trains and the Glendale plant usually made heavy gauge metal toys and large-size toys. (For detailed information about these plants, see Chapter 11 in *GREENBERG'S GUIDE TO MARX TOYS*, Volume I.)

In California, Mr. Wane says there were two bonded warehouses [not plants as has sometimes been thought]. One of the warehouses was in San Francisco. The other, in Los Angeles, was run by Marx relatives, jobbers who used the name of HOMEWOOD. Toys were stored in these warehouses and there would be no duty unless they were taken out.

Mr. Wane feels that there was no particular reason that Marx selected Pennsylvania for his main plants. The company just, "... fell into it. There was a vacancy and they took it."

Marx was able to buy the Glendale, West Virginia, plant for "... next to nothing because it was an abandoned Fokker Aircraft plant. This was during the Depression in the early 1930s and they just wanted to bring some employer in. Then they (Marx) subsequently built an addition, about seventy or eighty thousand square feet, which was a plastics plant." The plastics plant, having since been taken over by another tenant, is still running today.

CHARMORE AND OTHER MARX NAMES

Many collectors are familiar with the Marx use of the name LUMAR, frequently found on heavy duty trucks and road equipment, and also MARLINE seen on trains. Marx enthusiasts, being less familiar with CHARMORE may not realize that there is a Marx connection. Charmore was a jewelry line that was made by Marx after World War II. Toys were also made with the Charmore name. One of these with the very lengthy title, "Watch The Nursemaid Take Dick Tracy's Bonny Braids For A Stroll" is detailed in the Dick Tracy section of this book. The Charmore toy appears to be the same as the Marx version of the toy with the shorter title, "Nora The Nursemaid" except for a different box. Mr. Furie elucidates how the Charmore Company could be used effectively. "You take a buyer, you don't want to give him any money, so you say, 'Here's some stock in my company, Charmore.' Of course, when you are going over the toy line and you have ten items that are listed under Charmore, you know these ten items are going to be bought, don't you, if you're a stockholder in that company. And that was one of the ways you got ten more items bought by a customer, and then at the end of the year, I'm sure Charmore declared a dividend."

There was even one more company of which the author had never previously heard, CHARLMAR, named for Charlene, the wife of David Marx. Mr. Furie described Charlmar as having made clear lucite pocketbooks, very popular after World War II. All the various names were for different companies within the company. It was done, Mr. Furie gathered, for tax purposes.

MARX BUSINESS PRACTICES

As befits a major company, the Marx Company's business practices were many and varied. Of some surprise is the fact that the salesmen did not make deals in pricing. As Mr. Wane describes it, "When it came to pricing, everything was fixed and nobody had any leeway to do anything. We would set prices at our meetings, usually in December and the price structure was set and that was it." The salesmen did not have a bottom line against which they worked. "The salesmen worked hard, I'm going to give them credit. But they didn't work on commission, they worked on salaries."

There were no different prices for large quantity buyers. Usually, the small distributors did not buy directly from the Marx Company, they would buy through a jobber. "Jobbers would get a slightly different price, a lower price. However, if it was a plastic set, for example, there might be ten pieces fewer or twenty pieces fewer. Again, the price was the same as what we would charge Sears for the set with the larger number of pieces. The jobbers would get it for a little less, but they never knew that there were fewer pieces, and that's the way the business runs." Salesmen did not know the actual cost of a toy. They would help set the retail, the selling price, but they wouldn't know the actual cost of an item. Marx's profit margin varied depending on the item.

One of the Marx policies was competitive shopping. This job, among others, fell to Ed Furie. "I would go out and buy just about every known toy that was new, so we could evaluate them ... I probably spent about $12,000 a year. Everytime a competitor would come out with a truck that had a new cab, or a new style, or a new wheel, we would buy a sample and then evaluate it. The toy would be sent down to our plant in Glendale, or in Erie or in Girard. We always copied. We always bought electric trains from Lionel and Gilbert to see what they were showing, then we worked on it to see how we could copy it, how we could improve it. For example, a competitor would have a toy consisting of a rake, a hoe, a shovel and a cart - my job was to get that sample as fast as possible and send it down to the factory and we would cost it out. To give an example, we had a competitor called All Metal Toys (Wyandotte) that made a metal shovel that sold for ten cents. We made a shovel of the same size that sold for fifteen cents ... they (Marx personnel) costed it out and they couldn't figure out how All Metal Toys sold it for ten cents, just impossible. Eventually, All Metal Toys went out of business and Louis Marx bought them, in the late 1950s. The Marx Company took all the tools and sent them down to a plant we had in Mexico that was manufacturing for us, Plastimarx." When Marx Company personnel pulled out the cost sheets for the All Metal Toys shovel, "... I don't remember what the numbers were, but I always remember the outcome. It was, 'My God, it's costing them fifteen cents to manufacture and they're selling it for a dime!' You see, we never thought of it that way, it was costing us probably twelve cents to manufacture and we were selling it for fifteen cents. It cost them fifteen cents to make and they were selling it for a dime, just to get all the business. And they were getting all the business!"

Mr. Furie also travelled for the Marx Company, a policy the company started around the 1970s. Joe Kling, who was in charge of the Woolworth account, made the first road trip and then Mr. Furie was sent out, the second person to ever travel for the company.

In not having travelling salesmen until the later years, the Marx Company differed from other companies. However, it was thought unnecessary at the time because, "... people come in, they were qualified to buy and they wrote their orders at that moment."

Mr. Furie felt that the Marx Company was the most feared company in the industry. "Nobody would knock us off even though we would knock them off. They were scared. In fact, there was a firm called WOLVERINE which made toy stoves, sinks and refrigerators. We made the same things and we set prices for them. They used to come and ask, 'Well, how much are you going to sell your stoves, sinks, refrigerators and doll houses for?' We told them that we have the department stores and chain stores, they could sell to all the jobbers. In a way, we set their prices which were always higher than ours. In those days, a department store or chain wouldn't buy from them because they were higher priced and they knew that if they didn't do it that way, we would just annihilate them."

Basically, Mr. Furie felt the Marx Company treated others fairly. "We were good to everybody. It didn't make a difference if you bought $1,000 or $500,000. In those days, a million dollars was a big order. We treated everybody equally. The problem was that when we ran short of an item, the big guys got it first and the little guys got it second, if there was a chance to get it second. When we were running that BIG WHEEL, we were hot and could not supply all the demands. But we tried to make it up."

HEALTH INSURANCE

The Marx Company's dealings with the competition has been covered, but how did the company deal with its employees? There were pension contributions made by the company for the employees based on their salaries. The Marx Company, ever thrifty and with input from Archie Marcus, figured out a financially creative way to implement this plan at a low cost. This is how the plan worked, as explained by Ed Furie, "You came in for a job and they told you they were going to pay you $10,000 a year. You received a salary of $100 a week, or $5,200 yearly, and the other $4,800 was paid as a bonus at the end of the year. Well, it didn't make any difference in the $10,000 total, so you struggled the whole year on your $100 a week and then, at the end of the year, you went in there to get your bonus. And at that time you might even get a little raise, so that your $100 a week became $110. They gave you the $4,800 bonus and you ran around and paid off all your debts. Now the following year, during which you were being paid $110 a week, you went in there and they said, 'Well, we had a good year. We're going to raise you to $115 a week and raise your bonus from $4,800 to $5,000.'

"Now they were putting in contributions to your retirement fund based on your salary only, not including your bonus. But when you went back in to ask for a raise, they did count the bonus and said, 'What are you talking about? You're not making only $110 a week, what about the bonus?' Then they threw it back in that you were making $135.00 a week [based on the total compensation of weekly salary plus annual bonus].

"Being young and naive, the employees did not study the plan and when the first person to retire, an employee who had been with the company for thirty years, went to collect his retirement fund, he found out that he had very little in it. So, we had some people making $50,000 a year, but it would be $20,000 in salary and $30,000 in bonus.

"But we had nice bonuses. Nobody knew what was going on because employees didn't have to contribute [to the pension fund]. They put in whatever they wanted and they put in only on your salary, not your bonus, so they tried to keep the salary low." Mr. Wane adds that employees did have Blue Cross and Blue Shield insurance. The assembly line workers received strictly whatever the union negotiated.

DAVID MARX

Much has been written about Louis Marx, but comparatively little about his brother, David. Ed Furie, who was David Marx's assistant, knew this somewhat eccentric gentleman well. Mr. Furie feels that David Marx was not as well known as his brother, Louis, because, "... the company was named Louis Marx & Company, that was the first thing, and David Marx really only headed the finances. He was pretty good in the stock market and, of course, he had many friends, so he always got new issues of stocks. His job was really to pay the salaries and deal with the sales force, and very rarely did he deal with the buyers. He may have become involved with Woolworth buyers or K-Mart buyers (which in those days was S. S. Kresge) but he did not bother too much with running the business. It was Louis Marx who was in charge of the factories and it was Louis who decided what to make." Louis Marx did not have much to do with the sales force, since his brother ran that division and kept a tight grip on it. There are indications that Louis Marx would like to have spoken more with the sales force, but since David Marx paid them, he felt that they should report to him.

Mr. Furie has many recollections of David Marx. "He always liked new money. I used to have to go to the bank with a check, perhaps for a thousand dollars, and I would ask for $500 in tens, $200 in twentys and the rest in singles – all in new money. That bank used to go berserk, but I presume he had a lot of money deposited in that bank and he used to carry only new money.

"Another thing, if you had a cold, or if you coughed or sneezed, he would send you home and tell you not to come back until you cleared up your cold. I would say, 'Mr. Marx, I don't have a cold, it is my hay fever.' 'Don't argue with me' he would say, 'you have a cold.' I would say, 'Mr. Marx, if I go home, I'll never come back, it's hay fever time.'

"Dave was given an order once and he said to the buyer, 'What is this?' The buyer answered, 'That's my order.' As Dave was talking to the buyer, he was tearing up the paper into little pieces saying, 'You've made a terrible mistake, this couldn't be your order.' He just left all the pieces there, you couldn't put them together, there were thousands of pieces. And the buyer said to me, 'What do I do now?' I said, 'I presume you rewrite the order, but write it bigger than it was.'

"Or he would take a buyer's hat, saying, 'Where did you get such an ugly-looking hat?' and he would fling it, sometimes forgetting that in the summertime the windows were open. The hat would sail right out of the window and just as it left the window (he was very quick) he would reach in his pocket and grab whatever came into his hand, a fifty or hundred dollar bill, and give it to the buyer saying, 'Now, go buy yourself a good hat.'

"He used to tell me to go down and get his cough drops. In those days, there used to be maybe seven different brands, not as many as there are today. Well, I'd buy every brand and put them in my pocket and then when I'd hand him one, let's say Ludens, he would tell me that he never eats Ludens. I would say, 'Those are mine.' Then I would take out the next brand, which might be Smith Brothers, and he would take it. Next week, he might take the Ludens brand from me.

"I had a desk behind him. Once I moved away and he went berserk. He told me that I had to move my desk back, he wanted me behind him. You had to listen carefully to what he was saying because he never turned around to you. He would be sitting there reading the paper and, all of a sudden, something would hit his eye and he would say, 'Get me Doctor Kirk or get me Hank Greenberg (he was a personal friend of the baseball player).' He would talk really low and you had to be geared to him all day long." On one particular occasion, Mr. Furie found himself with nothing to do. David Marx said, "*Here's THE POST*, here's *THE WALL STREET JOURNAL*, here's *THE NEW YORK TIMES*, here's *THE DAILY NEWS*, just sit there and read."

"You never argued with David Marx. You never said 'but' and if you were right, you were still wrong, and when you were wrong, you were wrong. But then he would bring you back into his office, put his arms around you and say, 'Eddie, why do you make me yell at you? Here's a hundred, buy yourself something.' But everyone knew you were going to get a gift: twenty, fifty, whatever. You see, he always put his hand in his pocket and whatever he pulled out, he'd give it to you."

Such experiences might have been daunting to a lesser man, but Ed Furie was a man who was able to see the humor in a situation. And what did Mr. Furie think of David Marx? "Basically, he was a very nice person. His kids were nice, His wife, Charlene, was a wonderful lady to be around."

Mr. Furie liked his work at the Marx Company. "With all the yelling and other things, the company had loyalty to every employee and every employee was loyal to the company. It was fun when you were working there. You did everything, you learned a lot, they taught you the industry. You knew your competition, you knew all your buyers. The learning was wonderful. I knew every competitive item that was out there, just about. I doubt if I didn't know them all. In fact, I even had competitors call me up to ask if I could help them when they were trying to find out who had samples similar to theirs. Yes, Marx was just an exciting place to work."[3]

This chapter cannot be ended without a few words about Louis Marx, the man behind the Marx Company. Although detailed information is found in *GREENBERG'S GUIDE TO MARX TOYS*, Volume I, a couple of recollections from former Marx Research and Development worker Raymond Hubiak show certain facets of Marx's character. "I was quite impressed with the company. Here it was in the middle 1930s Depression period and Mr. Marx had the courage to move ahead with his enterprise, giving employment to hundreds of local area people throughout this rough period. We all appreciated it very much. Jobs were hard to come by."

Mr. Hubiak worked at the Girard plant for over twenty-six years. He was one of the few employees to meet Louis Marx twice when Marx made his infrequent visits to the plants. "I remember what he said on one of the trips. It concerned a problem we were trying to solve on a new or redesigned toy. He looked at me and said, 'Don't give up, you will find five or six ways of doing it' and he smiled. Quite a guy! With his large organization, he accomplished the 'Great American Dream.'"[4]

FOOTNOTES

1. Essential information from Roy Ahlgren was obtained on 1/17/89.
2. Seymour L. Wane contributed greatly to this chapter on 1/16/89.
3. Much of the information in this chapter was given by Ed Furie on 9/20/90.
4. Raymond Hubiak also contributed information on 2/01/89.

Robots and Other Space Toys

INTRODUCTION

Robots may seem to be a relatively recent phenomenon, but actually they were mentioned as early as 1921. At that time, Czechoslovakian playwright Karel Capek introduced the name "Robot," derived from the word "Robata" (meaning forced labor) in his play *ROSSUM'S ROBOTS*.

Then, in 1926, German director Fritz Lang's marvelous film *METROPOLIS* featured a wicked female robot. Interest in robots continued and after World War II it accelerated.

The climate that would create the impetus for the increased interest in robots and space toys is well described in the November 1989 *CHRISTIE'S CATALOG* of the Davidson Collection of Robots and Space Toys (no page number). The catalog states, "Post-war America's optimistic climate anticipated a better future through technology. Vehicles for space travel exploration were being developed and by the end of the decade, the first satellites were launched. Man had finally embarked upon his long-time quest to conquer the stars. In this future-oriented atmosphere, toys were produced based on the popular [1950s] films *WAR OF THE WORLDS, THE DAY THE EARTH STOOD STILL* and *FORBIDDEN PLANET*. [FORBIDDEN PLANET was particularly influential because of the popularity of the robot "Robby."] The centre of production for many of these fantastic toys was Japan, where technology was combined with artistry to produce toys..."

While some Marx Toys robots and space toys were made in the United States, the majority were made by Linemar, Marx's Japanese subsidiary. In addition, Lloyd Ralston, the well-known auctioneer and toy authority, and his knowledgeable son, Glenn, relate the intriguing information that, "We highly prize American-made robots because of scarcity and national pride. American robots were not exported to other countries the way Japanese robots were. They are hard to find here and nonexistent in other countries."[1]

Technological advances also affected the toys themselves. In his book *YESTERDAY'S TOYS,* Teruhisa Kitahara writes that, "even methods of printing on tin changed the world of toy robots, allowing the tiniest gauges and meters to be printed in detail."[2] One glance at space toys in general, not just robots, shows that such an advance was quite important.

Space toys are symbols of a technological age when it was thought that science would find the solutions to our problems. While this has not happened, it still has not lessened the desirability of these toys.

Adding to the attractiveness of the toys is the tremendous creativity embodied in them. Discussing robots in an article in the magazine *COLLECTORS' SHOWCASE*, collector H. I. Gosses explains, "Unlike trains, cars, bears and soldiers, they are not miniatures of real objects, but independent creations. Therefore, they can have an endless variety of shapes and functions. Without having to 'resemble' [reality] there is all the more opportunity to be beautiful and intriguing."[3]

The above certainly holds true for Marx space figures. They range in size from 1 1/2" to 38" high. They smoke, have lights and make buzzing noises, In one toy, the Hi-Bouncer Moon Scout, balls shoot forth from his chest, while in another, Big Loo, rockets are launched from his base.

While robots and some space figures are independent creations, this is less true of certain space vehicles. After all, there are such things as real satellites and space ships, but even here there is some leeway for creativity. For instance, what a flying saucer looks like is up to the imagination of the toy designer. The Marx Company and its Japanese subsidiary, Linemar, made some outstanding space toys and while not inexpensive, these underrated toys are not as much in demand as the robots and astronaut figures.

A factor adding to the current price of space toys is that many were battery-operated. While allowing the toy to perform all kinds of actions, the batteries caused problems. To quote H. I. Gosses again, "These automata devoured amps and soon the batteries would start to leak. A Lehmann car with a broken spring is still quite a good toy, but a robot that is oozing sulfuric acid ... in the ashcan!"[4]

For those avid collectors who covet the more expensive robots, but cannot afford them, non-moving ceramic copies of certain well-known robots are available, as well as metal copies with wind-up motors.

BIG LOO #7650: 1963. 38" high, battery-operated. Rare and handsome, this plastic toy is the tallest of the Marx robots.

Big Loo was introduced in the November 1963 issue of *PLAYTHINGS*. He was described as a versatile free-rolling toy "... reportedly able to bend, swing his right arm and pick up things in his spring-clasp right hand, fire four plastic balls out of his left forearm, launch soft-tipped rockets from his base, shoot two darts from his chest and squirt water.

"He comes complete with other distinctive features including a hand-crank record player with 10 sayings; blinking [red] battery-powered eyes (batteries not included); a built-in musical whistle; and a clicker signal. The toy also has a scanner target scope for spotting the 'enemy' and a mounted compass for tracking him."[5] Just what was the cost of this extraordinary toy? Considering the price that the toy fetches today, it sold for the puny price of $9.99 in the 1963 *SEARS CHRISTMAS CATALOG* (page 144). Read it and weep.

Big Loo has interesting coloring. The upper part of his body is green, the lower part burgundy. The entire body has gold trim. The top of his swiveling head is somewhat bullet-shaped. The back of his head has a crank which, when turned, enables him to speak his various phrases, such as, "I am Big Loo" and "Big Loo fights for you." The Morse Code is shown on his back.

The box is even harder to find than the toy. Made in similar colors to the toy, it shows the full-length figure of Big Loo and one of his arms holding a dart. It reads, "Your friend from the Moon, Big Loo." Indeed, the big smile on the face of Big Loo echoes the sentiments on the box. The other side of the box shows a larger Big Loo with his various actions listed on his body.

The name "Big Loo" was a good choice for the toy. Although spelled differently, it is quite probable that Big Loo was named for Louis Marx. Such a practice was not unknown with Marx toys.[6] Also, the double letter O O in L O O nicely matched the same letters in M O O N. A complicated toy such as Big Loo is difficult to find in good enough condition for all his functions to work.

Excellent $1,000 - $1500
Mint in Box $1,800 - $2,500

MARX BRINGS OUT GIANT ROBOT THAT WORKS, FIGHTS AND TALKS

Louis Marx & Company, 200 Fifth Avenue, New York City, is putting heavy Christmas TV support behind a giant moon robot that cleverly works, fights with various weapons, and even talks.

Called Big Loo (#7650), the robot stands more than 3 feet tall and is con-

structed of tough high-impact plastic. A versatile, free-rolling toy, Big Loo is reportedly able to bend, swing his right arm and pick up things in his spring-clasp right hand, fire four plastic balls out of his left forearm, launch soft-tipped rockets from his base, shoot two darts from his chest and squirt water.

He comes complete with other distinctive features including a hand-crank record player with 10 sayings; blinking battery-powered eyes (batteries not included); a built-in musical whistle; and a clicker signal. The toy also has a scanner target scope for spotting the "enemy" and a mounted compass for tracking him. Each robot is packed in an 8-pound shipping carton.

Marx Big Loo #7650 38" high, battery operated robot. Playthings. November 1963, p. 85.

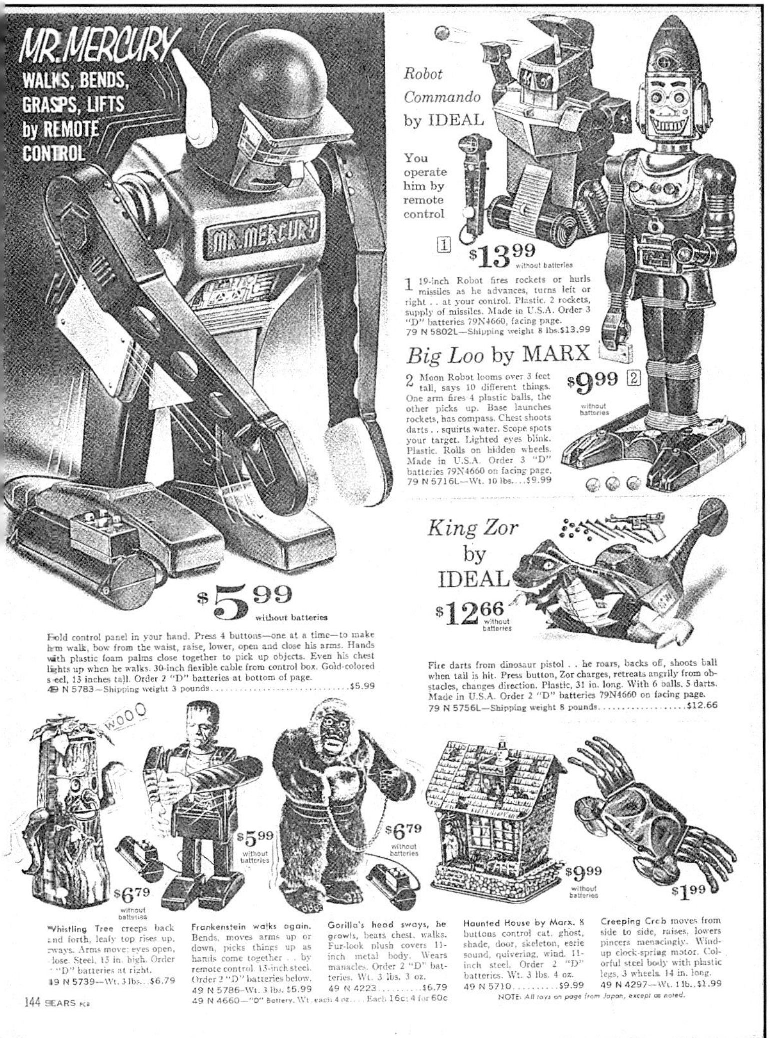

Big Loo as advertised in the 1963 Sears Catalog, p. 144. Several other Marx toys (including the Mr. Mercury robot described later in this chapter) are illustrated as well.

COLONEL HAP HAZARD: 1968. 11 1/2" high. Battery-powered.

Made by Marx Japan, Colonel Hap Hazard wears a white space suit with a red, white and blue NASA insignia on the chest, a red belt and suspenders, and black boots. The toy's body is metal except for plastic arms and rubber-like black plastic gloves. The toy's facial features are two-dimensional, but done in nice detail. A movable, clear plastic shield from his helmet covers his face. Atop Colonel Hap Hazard's helmet is a whirling blade which is lit with a red light on one end, a green light on the other. This blade is frequently missing from the toy.

On the figure's back are two red, plastic oxygen tanks. Below these are the Marx logo, the "Marx" name, the date "1968" and "Made in Japan."

The Colonel walks forward while his arms swing and the blade on top of his helmet whirls around.

The box reads "Whirling Lighted Blade/Forward Movement/Arm Movement." The box illustration shows the complete figure of the Colonel, blade whirling, while he stands within a lunar landscape.

Colonel Hap Hazard has the same body as another Marx toy, the Hi-Bouncer Moon Scout (see below), but the lithography of the two toys differs. If you like one of these attractive toys, you will like the other as well. Both toys together make a nice pair.

Excellent with box $1,200 - $1,500

COLONEL HAP HAZARD PROTOTYPE: 1960s. 11 1/2" high. Battery-powered.

As described in the November 1989 *CHRISTIE'S CATALOG* of the Davidson Collection of Robots and Space Toys, the prototype is, "Handpainted, including facial features, belts and NASA insignia, with Japanese notes on bottom, comes in Colonel Hap Hazard regular production box."

The prototype and production model are almost identical. However, the prototype has a different expression, appearing rather concerned, or even worried, as compared with the production toy's expression of steely determination. Also, the brim of the prototype's helmet follows the line of his eyebrows rather than dipping to a peak as in the production toy.

Prototype excellent with box $1,500 - $2,500

DALEKS: 1965-1974. 6 1/2" high. Battery-operated and Friction-powered versions.

Daleks are plastic robots from the British television series DOCTOR WHO. Battery-operated or friction-powered, they move with bump and go action. Large embossed knobs cover the body of the Dalek, topped by two gold bands. The "head" area has blue plastic windows through which a blinking light can be seen. The delicate and easily-broken arm and head came unassembled and need to be inserted. On the base is the embossed English Marx logo which reads "Marx Toys" over

The battery-powered Colonel Hap Hazard measures 11 1/2" high. He walks forward while his arms move and the blade on his helmet whirls around. Courtesy New England Auction Gallery, W. Peabody, Massachusetts.

the X surrounded by a circle. Around the circle is "Made in Gt. Britain."

Marx was not the only manufacturer to make Daleks. The British firm of Palitoy made a battery-powered talking Dalek, while another English firm, Condeg, made a mechanical Dalek.

DALEK: 1960s-1970s. 6 1/2" high. Battery-operated.

The battery-operated Dalek comes in a number of colors, such as a red body with red attachments and a black base, and a yellow body with yellow attachments and a black base.

The multicolored box for the battery-operated Dalek shows photos of other color variations, such as yellow with black attachments and a black base, and red with black attachments and a blue base. The box wisely states, "Colour of contents may differ from that illustrated." Other wording includes, "Marx Toys/Robot Action/Dalek/Battery Operated/Flashing Light/Uses 2 HP 11 Batteries." "Not suitable for children under three years of age due to small parts" and "From the BBC TV Series Dr. Who BBC 1974." In addition to the above color variations, it is known that the battery-operated Dalek also came in silver and black.

Battery version: Excellent with box $200 - $225

DALEK: 1965. 6 1/2" high. Friction-powered.

The friction-powered Dalek is similar in appearance to the battery-operated Dalek but far more difficult to obtain than the other version. Unlike the battery-operated version, the friction-powered version makes a siren sound. It is known to have come only in silver, although it is possible that it was made in other colors.

The friction-powered Dalek came in a different box than the battery-operated robot. Wording on the illustrated box is similar to the other toy but reads instead, "Friction drive with siren and flashing light."

Friction version: Excellent with box $250 - $450

DALEK CONSTRUCTION KIT: 1965.

The Dalek Construction Kit contains pieces of gray plastic which can be assembled into a Dalek of approximately 6 1/2" - 7" high, resembling those previously described.

The box for the kit shows the completed Dalek standing in the midst of a lunar landscape. In the distance, another Dalek can be seen in the air. The box reads, "The Dalek Construction Kit" and "From the BBC Series Dr. Who." The word "Dalek" is designed in the same interesting angular style as on

The 6 1/2" high battery-operated Dalek was made in the 1960s-70s. It has a bump and go action and a flashing light in its head. Shown with the Dalek Rolykin which measures only 1 1/2" high and rolls on a ball bearing. Samuel Caley Collection, Robert Grubb Photograph.

the boxes of the battery-operated and friction-driven Daleks. The Dalek Construction Kit version is even rarer than the friction Dalek.

Dalek Construction Kit: Excellent with box $250 - $450

DALEK ROLYKIN: 1965. 1 1/2" high.

The cute little Dalek Rolykin is a perfect miniature of its taller cousins. It is gray with red attachments, but it also comes in black or red.

Apart from size, what makes this Dalek different from the others is the steel ball bearing in its base. This ball bearing enables the toy to glide across a smooth surface.

The box for this Dalek is carefully illustrated, which is surprising for such a small and (at one time) inexpensive toy. Two sides show the Dalek rolling down an incline while the other two sides portray the toy in the sky among the stars. Wording and lettering are similar to the other boxes, except for "Dr. Who and the Dalek Rolykins."

As can be seen, Dalek items were quite popular. Marx even manufactured a Dalek Bagatelle Game about which no details are available.

Value: $75 - $100

(See battery-operated Dalek for photograph. Samuel Caley Collection, Robert Grubb Photograph.)

ELECTRIC ROBOT AND SON: 1950s. 14 1/2" high. Battery-operated.

Made in the United States, the plastic Electric Robot was originally sold without the Son in the early 1950's. In an amusing touch, a 5" high Son was added in 1957.

The toy comes in two color variations: the more common Robot is red and black with the Son all red in a silver diaper, while the Robot in the rarer version is maroon and silver with gold trim, and the Son is maroon in a gold diaper.

As well described by Glenn and Lloyd Ralston, "The robot is controlled by a lever on his back and moves either forward or backward. As he moves, his head turns from side to side and his eyes light up. His arms are controlled by knobs on his back and they can be adjusted up and down. On the back of his head he has a knob to adjust the height of his antenna [up to 4"]. Also, printed on the back of his head is a Morse Code chart. In the center of his back is a button connected to a buzzer for sending Morse Code. In a rectangular drawer above his belt there are tools for making repairs [hammer and screwdriver] ... the 5" Son is an absurd little robot (in a diaper no less) hanging from his parent's powerful pincer. This smiling baby robot has no action - he just hangs while his parent beams with pride."[7]

The box for the Electric Robot and Son is a wordy affair, listing all the various attributes of the toy. It is the same box as that for the Electric Robot but with the addition of the Son. In the foreground, the smiling Robot stands proudly with his happy Son. In the background, on either side of the toy, is a sparse-looking bush, which could just as well have been omitted.

The box reads, "Electric Robot and Son/Adjustable Antenna/The Eyes Light Up/Will Turn Right Will Turn Left/ Has Buzzer for Sending Morse Code. The antenna will adjust up to 4". The eyes are made of white light bulbs, which also blink. A directional switch near the back of the toy enables the robot to move left or right." The box's graphics include the 1950s logo and the well-known Marx slogan, "One of the many Marx Toys/Have you ALL of them?"

Value: Electric Robot with box $300 - $400
Electric Robot and Son with box $450 - $475

The Electric Robot and Son were manufactured in the 1950's. Robot is 14 1/2" high. He moves backward and forward, his head and arms also move and his eyes light up. Son is 5" high and does nothing but look cute. Courtesy LLOYD RALSTON GALLERIES, Fairfield, Connecticut.

ELECTRIC REMOTE CONTROL BATTERY-OPERATED ROBOT (R-35): 1950s. 8" high.

This very desirable tin robot is commonly known as "R-35" because of the markings on his back, or "Robot Control" because of the words on his battery box. The actual name, as listed on the box the toy came in, is "Electric Remote Control Battery-Operated Robot."

The toy was marketed under the name of Marx's Japanese subsidiary, Linemar, and also under the name of the Japanese manufacturer Masudaya. (For more on Linemar, see the Introduction to this book.)

The robot is either gray with blue eyes or blue with green eyes, while the battery box can be either a lithographed, illustrated tin version (the one seen most often) or the rarer unillustrated plastic version. Both robot color variations have been seen with tin battery boxes, but only the blue version has been seen with the plastic battery box. However, it is probable that the gray version also came with the plastic battery box.

The most noticeable feature of the robot is the outstanding lithography of gauges and meters on his chest. Above these, a highly stylized red "Robot" is written with the bottom edge of the "R" extending in a line to the bottom of the "T." There is a simulated rivet design over the body. The toy's head is bucket-shaped while his nose is triangular. The glass eyes light up and blink, and he can move forward or backward, while his arms swing.

The lithographed tin battery box is illustrated with three robots of varying sizes who march forward while rays of light shine from their eyes. The light of the largest robot is so exaggerated that it extends from almost one side of the battery box to the other. The plastic battery box is plain and unillustrated.

The box for the robot shows a side view of the blue version and lists all his various activities. The red word "Robot" is amusingly designed as if made out of gears and rivets.

The robot is so popular that reproductions have been made of him. One by Masudaya is half the size but is mostly tin with plastic parts and has a wind-up motor. Another reproduction is a few inches taller than the Linemar toy, but it is made of ceramic and does not move.

Robot with tin battery box: Without outer box, excellent $595 - $895.
Robot with tin battery box: With outer box, excellent $1,000 - $1,200.
Robot with plastic battery box: $200-$300

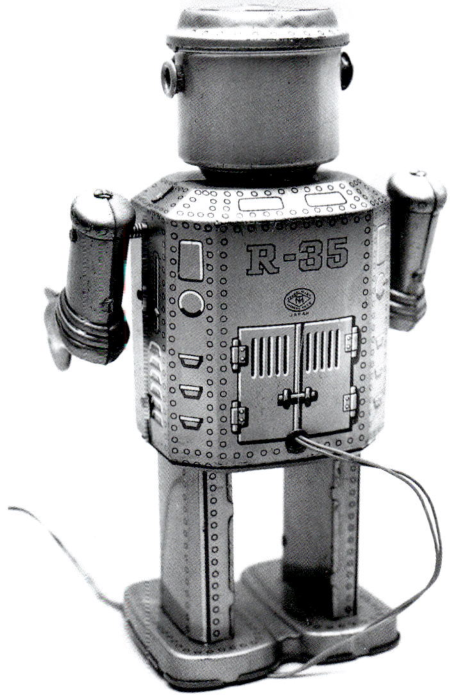

Side and back view of R-35 Robot. Courtesy LLOYD RALSTON GALLERIES, Fairfield, Connecticut.

Robot popularly known as R-35, or Robot Control, measures 8" high and was manufactured in the 1950s. It was marked Linemar or Masudaya. The toy moves forward and backward while his arms swing and his eyes light up and blink. Courtesy SMITH HOUSE, Eliot, Maine.

GOLDEN ROBOT: See Robot, Golden.

HI-BOUNCER MOON SCOUT: 1968. 11 1/2" high. Battery-powered.

Although colored differently, the Hi-Bouncer Moon Scout has the same body as Colonel Hap Hazard. He is marked in the same manner with the Marx logo, name and date on his back, along with "Made in Japan." Unlike the Colonel, the Hi-Bouncer Moon Scout is remote-controlled. As described by Glenn and Lloyd Ralston, "This astronaut is silver with a metal body and helmet. His helmet is mustard colored and has the same flip-down, clear plastic visor and antenna as Colonel Hap Hazard. The face of this astronaut is slightly off center, but at least it is a lot friendlier [than that of Colonel Hap Hazard]. Lithographed on the center of his chest are an American flag and two zippers. On his back he has two white plastic air tanks. His arms are silver colored plastic with light blue gloves. The legs are also silver plastic with light blue tin boots."[8]

The Hi-Bouncer Moon Scout, like Colonel Hap Hazard, walks forward while his arms move and the blade on his helmet spins. However, there is a dramatic difference between the two toys, for in the Hi-Bouncer Moon Scout's chest is a trap-door which opens, shooting out super balls, one at a time. The balls are loaded through an opening in the back of the helmet.

The box for the toy lists all the various features of the Hi-Bouncer Moon Scout. The astronaut is illustrated marching intrepidly forward while his blade whirls and balls fly from the open door in his chest. The Hi-Bouncer Moon Scout is harder to find than Colonel Hap Hazard.

Value: Excellent with box $2,000 - $3,000

JANE AND JOHNNY APOLLO ASTRONAUTS: 1968. 7 1/4" high, each. Non-powered.

These non-motorized poly-plastic astronauts have fully articulated arms and legs. They can bend at the waist and their wrists can be moved completely around, while their heads can turn from side to side. The pair have flesh-colored skin and wear detailed astonaut suits with NASA emblems, with the Marx logo embossed on their backs.

Jane Apollo has blonde hair, blue eyes, and wears bright red lipstick. She is dressed in a blue space suit. Johnny Apollo has dark brown hair, brown eyes, and is dressed in a white space suit. The toys have many accessories; those that came with Jane are white, while Johnny's accessories are gold. As listed on their boxes, they include "Helmet, Long-billed Cap, Leg Type Clipboard, Guidance Gun, Camera, Air Conditioning Unit, Back Pack, Nub Tool, Planet and Combination Tool." The toys also came with various decals and an equipment manual.

Collector Rick Johnson describes the box lid for Jane Apollo as having a photo of the toy and a background drawing featuring a launch pad at the Cape. However, as he alertly points out, the rocket depicted is actually an Atlas Booster from Project Mercury, not a Saturn V from Apollo.

Although the Johnny Apollo toy is not uncommon, the Jane Apollo toy is found less frequently.

The box for Johnny is a combination of photos, illustrations and printing. On the front is a full-length photo of the astronaut and in the background there is an illustration of a rocket being launched in a space center. The box reads "Johnny Apollo Astronaut Movable Spaceman by Marx/Fully Jointed! Made of Poly-Plastic." The photo of just the head and shoulders of the astronaut is repeated on the top and bottom of the

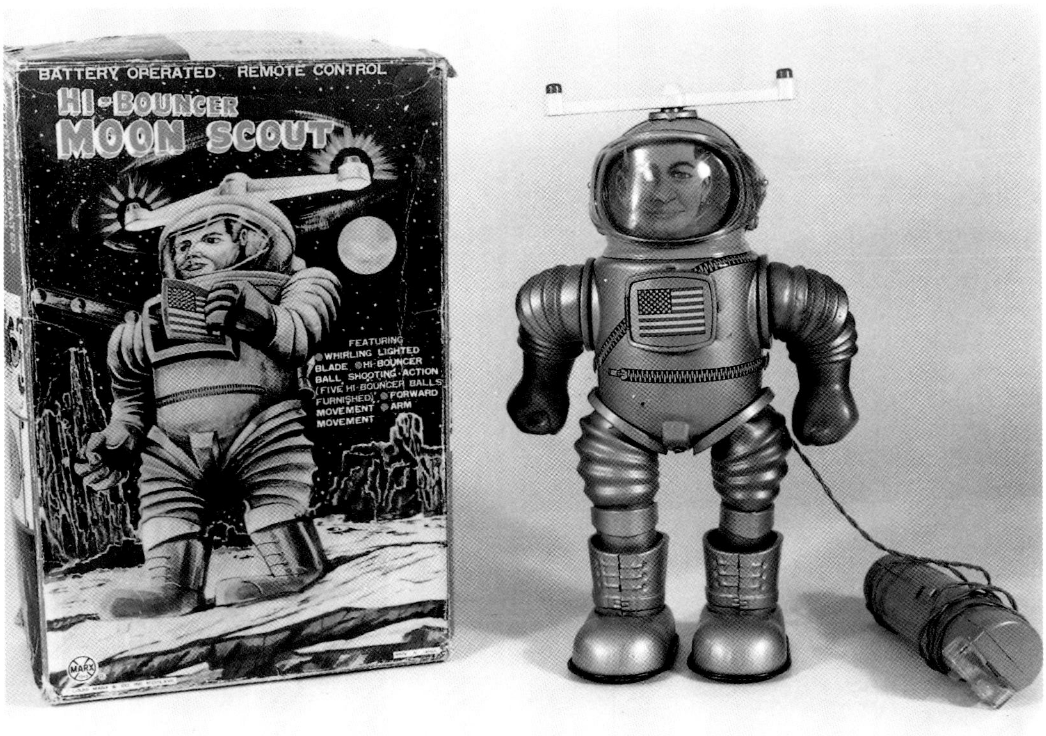

Hi-Bouncer Moon Scout is 11 1/2" tall and was manufactured in 1968. He walks forward, arms swinging, while the blade on top of his helmet spins. Super balls shoot out through a door in his chest. Courtesy LLOYD RALSTON GALLERIES, Fairfield, Connecticut.

box, along with "Louis Marx & Co., Inc. Glendale, W. VA. 26038." The sides of the box list the equipment that comes with the toy (see above) as well as the words "Johnny Apollo Astronaut/The Completely Movable And Fully Equipped Spaceman/Put Him In Different Poses." The box for Jane Apollo is probably almost identical except for her photograph and wording which refers to her.

Jane and Johnny Apollo were sold separately for $2.77 each in the 1968 *MONTGOMERY WARD CHRISTMAS CATALOG.* (Interestingly, the 1968 *SEARS CHRISTMAS CATALOG* did not show the toys.) Also sold with the astronauts were two space vehicles, the "Astro Buggy" and the "Moon Crawler." These could also be purchased separately. (See "Other Space Toys" in this chapter for further information on the two space vehicles.)

The Jane and Johnny Apollo dolls, along with the Astro Buggy and the Moon Crawler, also came as a complete set called Space Expedition. It sold for $9.99.

The Johnny Apollo Asronaut toy was so popular that Marx also manufactured a Johnny Apollo Bagatelle Game and a Johnny Apollo Space Pistol.

Value: Johnny Apollo $100-$150
Jane Apollo $125-$175.

MARK APOLLO ASTRONAUT: c. 1968. 7 1/4" high. Non-powered.

This toy is identical to Johnny Apollo Astronaut except that this astronaut wears a bright orange outfit and his accessories are white, as described in *HAKE'S AMERICANA & COLLECTIBLES CATALOG #111* for November 1990. Even the box is the same except for the photo and the name "Mark" instead of "Johnny."

Value: $100-$150

MECHANICAL WALKING ROBOT

This appealing and rare robot made by Linemar, Marx's Japanese subsidiary, is a mixture of blue tin and red plastic. His plastic legs walk with careful step-over action, as his plastic arms, which appear to be jointed but are not, move up and down. There are supports on his feet to help him balance. The crown of the toy's head is also plastic as are his eyes, nose and ears.

The robot's chest is decorated with various gauges and meters. Two oxygen tanks are lithographed on his back. Below these is the Linemar logo and "Japan/Linemar Co., Inc."

The robot comes in two variations.

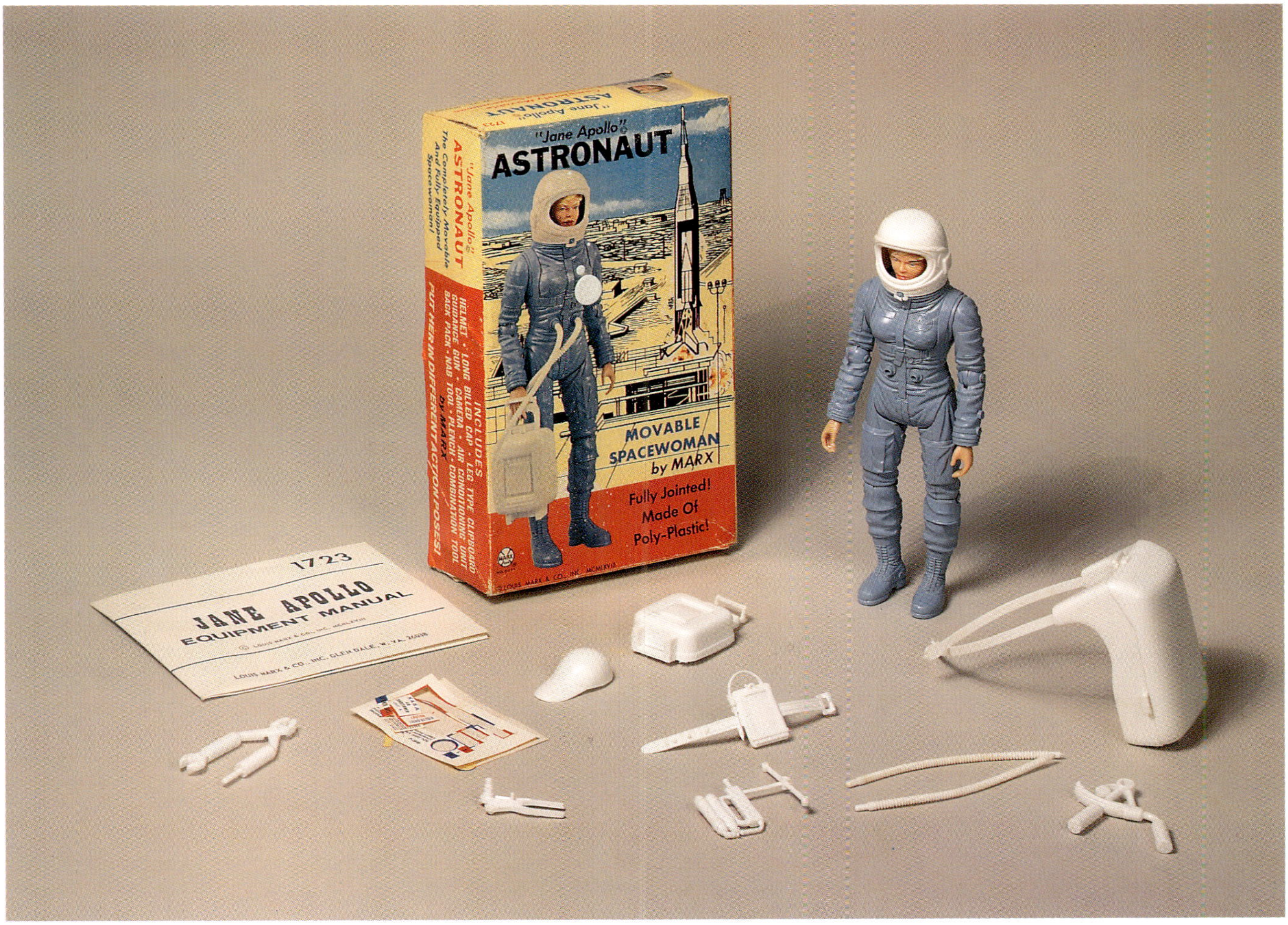

This 7 1/4" tall plastic Jane Apollo Astronaut was manufactured in 1968. Her body is fully movable. She is less easily found than her partner, Johnny Apollo. Rick Johnson collection, Robert Grubb photograph. The 7 1/4" tall plastic Johnny Apollo Astronaut, partner to Jane Apollo Astronaut, was manufactured in 1968. The toy is non-powered. He has movable limbs and head, and comes with ten pieces of equipment. Samuel Caey collection.

MECHANICAL WALKING ROBOT (A): 1950s. 6" high. Wind-up.

In this, the rarer variation of the robot, the word "ROBOT" appears below the chest, which slants out at the lower end. This variation has not been seen in action but it is believed that sparks are emitted from the toy's chest.

The box for this variation is illustrated with a rather masterful-looking robot in contrast to the somewhat boyish-looking actual toy. The robot is shown, chest sparking, standing within a lunar landscape. In the distance, a rocket stands poised for takeoff. At the bottom of the box is the Linemar name and the slogan "Best By Far."

Value: $225-350

This 1950s Mechanical Walking Robot measures 6" high. He walks while his arms move. The lower part of his chest slants outward and is believed to spark. It is marked "ROBOT" and is the rarer of the two variations. Courtesy LLOYD RALSTON GALLERIES, Fairfield, Connecticut.

MECHANICAL WALKING SPACE MAN (B): 1950s. 6" high. Wind-up.

This variation resembles (A) except for the following differences: the chest area is flat and does not spark, and the name below the chest is "SPACE MAN" instead of "ROBOT."

This toy comes in the same box as the "Robot" variation, but the wording merely reads "Mechanical Walking Space Man."

Value: Excellent $250-$350
Excellent with box $1,200-$1,500

This 1950s Mechanical Walking Space Man is similar to the variation marked "ROBOT" but is marked "SPACE MAN" instead, and the chest does not slant out or spark. Samuel Caley collection, Robert Grubb photograph.

MOON CREATURE: 1968. 6" tall. Wind-up.

The amusing Moon Creature is all tin except for his two vinyl head feelers, the plastic bubble over the button eyes (although these are ringed with metal), and the three rubber-like plastic wheels in his base.

The toy has horizontal ridges across his body. His head consists of a larger section over a smaller section. The larger section rises, making the toy appear as if he has a very large mouth. Lithographed on the Moon Creature's back are the Marx logo and "Louis Marx & Co., Inc./1968/ Made in Japan."

The Moon Creature rolls forward, arms moving, while his mouth opens and closes, accompanied by a clicking sound.

The toy comes in two color variations. In the first and more common, he is yellow and cream with blue trim. Red rings are around his neck and wrists. Large red lips are painted on his face.

In the second variation, the Moon Creature is white with a red belt and red suspenders. His face is flesh-colored and he lacks the big lips of the other variation and his nose is shaped differently. Strangely, although the second variation is rarer than the previous one, it can sometimes be bought for the same price.

Both versions were sold with the same box. The box front shows the common variation, standing on a lunar surface. On the sides of the box, both variations are illustrated. Designing the box in this manner was a wise move for it meant that the same box could be used for either version.

Value for either variation: Excellent with box $250-$275.

Moon Creature measures 6" high and was manufactured in 1968 by Marx Japan. When wound, the creature moves forward, arms swinging, while he opens and closes his mouth with a clicking sound. Courtesy NEW ENGLAND AUCTION GALLERY, W. Peabody, Massachusetts.

The second variation of the Moon Creature is rarer than the first and has different lithography. Samuel Caley collection, Robert Grubb photograph.

MOON INSECTS

There are two of these Moon Insects, and except for the lithography and certain small differences, they are the same as the Moon Creatures. It is safe to say that these insects have never been seen anywhere on the earth.

CHOMPY THE BEETLE: Circa 1968. 6" high. Wind-up.

The colorful Chompy is red, black and yellow, with blue hands and blue circles on his back. Around his mouth is a jagged red design probably meant to represent teeth. No nose is lithographed on his face.

Unlike the lithographed markings on the backs of the Moon Creatures, the Marx logo and "Louis Marx & Co., Inc., Made in Japan" are embossed on the base of the toy. No date is indicated this time.

The box front for Chompy reads "CHOMPY THE BEETLE" and shows him nibbling on a leaf. The side of the box shows two different color variations of the toy, one with a blue head and the other with blue arms and back, with the other coloring remaining the same. It is possible that these variations exist, but they have not been seen. Wording on the box side reads "CHOMPY THE BEETLE MOVES FORWARD WITH CHEWING ACTION AND SOUND." Also on the side is the Marx slogan, "One of the many Marx Toys, Do you have all of them?"

Value: Excellent with box $175-$200

Chompy The Beetle, made by Marx Japan, is a differently lithographed variation of the Moon Creature. Samuel Caley collection, Robert Grubb photograph.

INSECT "7": Circa 1968. Approximately 7" high. Wind-up.

Insect "7" resembles the Moon Creature in appearance. He is actually the same size except that his head feelers are straight with knobs on the tips, as opposed to the V-shape of the Moon Creature's feelers. Insect "7" (with the quotation marks) is lithographed on the toy's chest. Insect "7" is the rarest of all the variations of this toy.

Value: $100-$225

Insect "7", manufactured circa 1968 by Marx Japan and measuring 6"-7" high. Courtesy LLOYD RALSTON GALLERIES, Fairfield, Connecticut.

MR. MERCURY: 1962-1964. 13" high. Battery-operated.

The striking Mr. Mercury was manufactured by both Marx and its Japanese subsidiary Linemar. The latter version was also manufactured by the Japanese firm of Yonezawa. The Linemar version, which is believed to be the rarest, was advertised in the 1962 *SEARS CHRISTMAS CATALOG* (page 450) where it sold for the now tiny sum of $5.99. Mr. Mercury was described in the following way: "He walks ... bends ... grasps and lifts things – all by remote control. Amazing robot bows ... raises arms ... all at your command ... even closes hands to pick up objects. Whirring noise and slow precision add to realism. Head lights up to reveal figures of men behind clear plastic windows."

Lloyd and Glenn Ralston, who owned both variations, describe the Linemar Mr. Mercury as having a gray body with red plastic lettering across the chest and a red light mounted in a gray housing. The red light glows when Mr. Mercury is operating. The upper arms are blue and, from the elbow down, white [and red] plastic. There are sponge pads on the hand area [sometimes these are missing or in poor condition]. He has black legs with oversized orange feet to keep him from falling down when he bends over. His "eyes" are clear plastic and he has white plastic ears. His head (which is his control center) is blue with an orange cap, gray nose and mouthpiece. The robot is operated by people inside him. His pupils are two operators controlling him.

The Marx Mr. Mercury was advertised in the 1964 *SEARS CHRISTMAS CATALOG* (page 205) where it sold for $5.99, the same price as the Linemar version sold for two years earlier. The Marx version comes in two variations, neither of which have the red light, plastic arms or plastic lettering across the chest as in the Linemar version. In the Marx versions, the name "Mr. Mercury" is lithographed across the chests of the toys instead of being in red plastic. With the exception of the differences listed above, the first Marx variation is the same as the Linemar version.

The second variation has a gold body with gray arms, legs and head. His helmet and shoes are red, while the rest of his coloring remains the same.

It is interesting to note that Mr. Mercury's head is the same shape as the Talking Robot distributed by Cragstan, while his body (except for his hands) is basically the same shape as the Marx Frankenstein toy.

As the Ralstons point out, unlike many robots which have instructions printed on their boxes, Mr. Mercury came

The 13" tall Mr. Mercury was manufactured 1962-1964 by Marx and also by Linemar. The two Marx color variations are shown on either side of the Linemar version. Courtesy LLOYD RALSTON GALLERIES, Fairfield, Connecticut.

Boxes for the Marx and Linemar versions of Mr. Mercury. The Marx version was marketed in the box on the left. The other box, although marked Marx, appears to illustrate the Linemar version. Courtesy LLOYD RALSTON GALLERIES, Fairfield, Connecticut.

with a separate instruction sheet.[9] The box illustration for the Marx Mr. Mercury shows him lifting a large rock from a lunar surface. The box illustration for the Linemar Mr. Mercury shows him lifting what looks like a large box. In the background of the lunar landscape, another Mr. Mercury stands beside a rocket poised for takeoff. Both boxes list the various actions of the toy.

Value: Excellent $700-$1,000
Excellent with box $1,200-$1,500

MR. SMASH: 1970s. 5" tall. Wind-up.

A late and rather uncommon Marx toy, Mr. Smash was described *in HAKE'S AMERICANA & COLLECTIBLES CATALOG #110* for September 1990. The robot is silver with a large red head on which two thin plastic antennae can be inserted. It has protruding black eyes and a sticker on his chest marked "For Mash Get Smash." He has two claw-like arms and one of these appears to be holding a potato. The robot walks forward, opens and closes its mouth, and raises one arm (which arm is unknown).

The box for the toy reads "Mr. Smash/Clockwork Walking Smash Martian." It is illustrated with color photographs of Mr. Smash plus two different photos of what appear to be life-sized robots sitting around a table where a television commercial is being filmed.

Value: $100-$250

ROBOT: 1950s. 6" tall. Wind-up.

The toy known simply as "Robot" is tin and was made by Linemar, Marx's Japanese subsidiary. The toy was also marketed under the name of the Japanese manufacturer Yonezawa, who made two other robots that were similar in appearance to this one.

The robot is silver with red and black trim. On his stomach is a target-like circle with two lightning bolts, in the center of a three dimensional-appearing rectangle. Two different types of meters are shown on either side of the robot's chest, on the left side a circular meter, on the right side a rectangular meter. His arms are jointed at the shoulders, elbows and wrists.

The robot has circular eyes, a triangular nose and a rectangular mouth. A spring-shaped antenna runs horizontally across his head to his ears.

A single bolt of lightning is lithographed on the robot's back, while on his lower right leg is the Linemar logo and the word "Japan." Extending from his waist to the floor is a triangular-shaped wire which helps to support him as he walks.

The box for the toy reads simply "Mechanical Robot" with the word "Mechanical" much smaller than "Robot." The robot is illustrated walking forward. This box is difficult to find.

Value: Excellent $500-$550
Excellent with box $700-$750

This 6" tall robot was manufactured by Linemar in the 1950s. He walks forward with the help of a wire support in his back. Courtesy SMITH HOUSE, Eliot, Maine.

Back view of Robot showing the triangular-shaped wire that helps him walk. Courtesy LLOYD RALSTON GALLERIES, Fairfield, Connecticut.

ROBOT (GOLDEN): 1950s. 6" tall. Battery-operated.

Made by Linemar, Marx's Japanese subsidiary, this tin remote-controlled robot is popularly known as "Golden Robot" because of the golden color of his body. He is trimmed in red and his predominantly yellow chest shows various gauges and meters. Over his body is a lithographed design of silver rivets.

The Golden Robot has circular eyes, a triangular nose and wears a big smile. The top of his head is flat. The toy walks forward or backward while the arms move and eyes light up.

The illustration on the box varies slightly from the actual toy. The word "Robot" is shown on the toy's chest in the illustration, but does not appear on the actual toy. Another difference is that the arrangement of gauges and meters on the illustration varies from those on the toy itself.

Wording on the box reads in large letters "Robot," then in smaller letters "With Flashing Lighted Eyes/Moves Arms/Reverse and Forward/Remote Control/Battery Operated." The word "Golden" is not mentioned. The robot is shown walking forward, light shining from his eyes, while behind him various heavenly bodies and a satellite can be seen.

The Golden Robot is difficult to find and the box is even harder to locate.

Value: with box $4,025-$4,075.

ROBOT CAR WITH SPARKS: 1950s. 6" long. Friction.

This rare Linemar sparking toy consists of a robot (actually just its head) inside a car. The royal blue robot has a spring which reaches across its head and is attached at the ears. The spring attachment is similar to that on the "Robot" described earlier.

In front of the robot, on the right side, is a blue projectile. In back of the robot is a blue fin, similar to an airplane's rudder. The yellow and royal blue car is bullet-shaped in front with a straight back.

The original box reads "Robot Car With Sparks" and has the Linemar name and logo. The robot car is illustrated and in back of it is a rectangular picture of a lunar landscape with a black sky. The box is primarily designed in yellow and blue to match the toy.

Value: Excellent with box $1,000-$1,500

ROBOTRAC: 1950s. 9" long, 7" high. Battery-operated.

Robotrac was manufactured in the 1950s by Linemar. It measures 9" long and 7" high. When the bulldozer encounters an obstacle, the robot shifts gears and a horn sounds. Courtesy SMITH HOUSE TOYS, Eliot, Maine.

This impressive tin toy was manufactured by Linemar, Marx's Japanese subsidiary. It varies from those described previously in that this robot drives a bulldozer. The side of the bulldozer reads "Robotrac" but this name is not shown on the toy's box. The bulldozer is primarily red, yellow and orange with a design of simulated rivets.

The robot driver is silver with red, yellow, black and white trim. The front of his body is lithographed with various gauges and meters. Like the bulldozer, he also bears a rivet design. His arms and legs are jointed. Like the toy known simply as "Robot," a spring runs across his head to his ears. Also on his head is a red bulb that lights.

The bulldozer speeds along until it encounters an obstacle. Then, the robot, with arms and legs moving, shifts gears and a horn blows.

Thr box shows a front view of the toy with the light shining from the robot's head. Inside the stream of light are the words "Bulldozer With Horn And Light."

It is believed that Linemar did not manufacture any other toys of this type, which is a pity, although plenty of Japanese manufacturers did, including Yoshiya and Showa, as did the English manufacturer Saunders.

Value: Excellent with box $450-$900

ROBOT WITH LIGHTED EYES AND SMOKE AND LIGHTED LANTERN: 1950s. 8" high. Battery-operated.

This remote-controlled tin robot was made by Linemar, Marx's Japanese subsidiary. He is gray with lithographed silver rivets all over his body. Teruhisa Kitahara, in his book *ROBOTS/TIN TOY DREAMS,* shows the same robot, but with blue head and arms. Otherwise, the robots appear to be the same.[10]

On the robot's chest, there are red and silver designs consisting of two rectangles with vertical lines and, in the center, a circle with a lightning bolt. The robot's back is particularly interesting. As described by Glenn and Lloyd Ralston, there is a cutaway view in red, silver and black, showing gears and pulleys. Underneath is the Linemar logo and "Made in Japan." They further state that the robot's arms have the same rivet detail as the rest of the body and have chrome claws, one of which [the right] holds the bright red lantern.

The head of the robot is square [similar in shape to that of the Golden Robot] with the top opening having an antenna lithographed on it in red and silver. The eyes are bulbs surrounded by red trim and his ears and nose are red.

When activated, the robot lumbers forward with eyes lighted while the lantern on his arm sways. Lloyd and Glenn Ralston describe, with experience, the robot's smoke action: "The smoke (talc) is blown out of a hole in the mouth and if you operate the toy long enough, you can make quite a mess, as the talc does not dissipate the way the smoke from other smoking toys does. However, the fact that this robot uses talc for the smoke instead of the awful-smelling stuff used in the other smoking toys is great."

The robot came with two sets of instructions, they state. One set is for loading the battery box, while the other set deals with the loading of the smoke chamber. The latter is as follows: "1. Open the lid on top of the head. 2. Remove inner tin lid from the powder holder, fill the holder half full, replace the lid tightly and close head cap. 3. Ordinary talcum powder is used to refill the smoke unit. 4. If the orifice should stop up, use thin wire or pin for cleaning."[11]

The box for the robot is similar in design to that of the Golden Robot, also showing a stream of light shining from the robot's eyes with the name "ROBOT" highlighted in large letters. In smaller letters is "with lighted eyes and smoke and lighted lantern." Also on the box are "Remote Control/Battery Operated," the Linemar logo and slogan "Linemar Best By Far."

Value: ? ?

This 1950s robot was manufactured by Linemar and is 8" high. He carries a lantern in his right hand and, when activated, he walks, his eyes light and he blows smoke (talcum powder) through his mouth. Courtesy LLOYD RALSTON GALLERIES, Fairfield, Connecticut.

Robot showing open lid on top of head. Talcum powder is inserted in container inside head. Two different sets of instruction sheets are shown in front of toy. Courtesy LLOYD RALSTON GALLERIES, Fairfield, Connecticut.

ROCK 'EM SOCK 'EM ROBOTS: 1964. 20" across including controls, 12 1/2" high. Lever-controlled.

A very beloved toy, the Rock 'Em Sock 'Em Robots is known to both collectors and non-collectors, probably because many owned this popular Marx toy when they were children. People have difficulty standing near this toy without wanting to play with it. It was created by Marvin Glass, a free-lance designer who also offered his work to companies such as Ideal and others.[12]

The 1964 *SEARS CATALOG* (page 120) shows this toy priced at $7.66. The plastic toy is really a game consisting of two fighting robots in a boxing ring. The object of the game is for one robot to hit the other on the chin. When a robot is properly hit, his head shoots up on a metal rod and he utters a startling squeal. His head is reset by manually pushing it down into place. Oddly, no advertising literature seen mentions that the toy makes a sound.

The red and blue levers on each side of the boxing ring move the boxers from left to right, and buttons on top of the levers move the boxers' jointed arms up and out. The levers are connected to the boxers by steel rods that are visible on the inner sides of the robots' legs.

The sturdy-looking robots are red and turquoise blue with an embossed rivet design on the sides of their heads, arms and legs. It is not generally known, but they came with decals on their backs lettered "Red Rocker" and "Blue Bomber." These decals are usually missing and their absence should not be enough to deter a collector from buying the toy. It is even possible that some were sold without the decals. (The 1974 *MARX CATALOG* has a photo of the toy without the decals. Even though Marx made "Clash of the Cosmic Robots" in the 1970s, as described below, the original Rock 'Em Sock 'Em Robots was still being shown in the 1974 catalog.)

The boxing ring mat (floor) is yellow with an embossed design of diamonds surrounded by rectangles. On the side of the ring is a decal that reads 'ROCK 'EM SOCK 'EM ROBOTS BY MARX" and at each control is a decal with a warning "Do not handle arms at any time." The "ropes" for the ring consist of two strands of white vinyl secured by four removable yellow plastic pegs, one at each of the ring's corners.

Included in the box for the toy was a cardboard backdrop illustrating the audience attending the fight. Above the audience hangs a banner proclaiming "FIGHT NITE - CIVIC AUD." This backdrop adds to the toy's appeal, but like the robot name decals, it is sometimes missing. Also included in the toy's box were score cards to add up the rounds, another item not always found with the toy.

The box for the Rock 'Em Sock 'Em Robots shows an illustration of one robot hitting the other on the chin, while the other robot's head shoots up on its rod. Also on the box are the words "Rock 'Em Sock 'Em Robots by Marx/Knock His Block Off." The toy's title is repeated on the box's side along with "Round by round score sheets included." The illustration on this side shows two boys enthusiastically playing with the toy.

It is believed that in the 1970s, the Rock 'Em Sock 'Em Robots was made in a more streamlined version known as "Clash of the Cosmic Robots."

The Rock 'Em Sock 'Em Robots was also sold in England under the name "Raving Bonkers The Fighting Robots" ("bonkers" is British slang for crazy). This version was sold circa 1972, but resembled the 1964 version. The box has a photograph of two children operating the toy. The robots are shown "speaking" to each other by means of cartoon balloons. The robot on the left exclaims, "What a terrific right hook," while the one on the right says, "There goes my head again."

In the mid-1980s, Durham Industries manufactured a toy called "Battlin' Robots" which is very similar to the Rock 'Em Sock 'Em Robots. Although the colors of this toy (gray, black and red) are different and the robots vary to some degree (one robot more so than the other), the basic workings and size of the toy resemble the Rock 'Em Sock 'Em Robots. Even the design of the boxing ring floor is very similar.

So continuous is the popularity of the Rock 'Em Sock 'Em Robots toy that the manufacturer Remco has issued a new version called "Totally Awesome Dueling Dudes." It is basically similar to the original Marx version except that the boxers are named after two of the characters from the Teenage Mutant Ninja Turtles, "Michelangelo" and "Shredder."

To sum up, the Rock 'Em Sock 'Em Robots is not one of the most expensive robot toys. In fact, it is one of the most reasonably priced, but it is a fun toy.

Value: With box $100-$200

The popular Rock 'Em Sock 'Em Robots was manufactured by Marx in 1964 and measures 30" across and 12 1/2" high. When one robot boxer hits the other on the chin, the head of the boxer who is hit shoots up on a rod and he utters a startling sound. Battlin' Robots was manufactured by Durham Industries in the mid-1980s. It greatly resembles the earlier Marx Rock 'Em Sock 'Em Robots. Samuel Caley collection, Robert Grubb photograph.

Box for Rock 'Em Sock 'Em Robots. Courtesy Smith House, Eliot, Maine.

RIGHT: Elliot Handler, president of Mattel, points out company's billboard sign at Broadway and 46th to Luke Halpin, MGM-TV personality, and to Ruth Handler, executive v-p.

BELOW: In new Milton Bradley showroom at Fifth Avenue Building, marking the company's presentation, on TV, of "Shenanigans," starring Stubby Kaye, are (l-r): President James J. Shea; Mr. Kaye; James J. Shea, Jr., executive v-p; Charles A. Shea, v-p-sales; and adman Ken Carlson.

Here and There

LEFT: David Koffman, sales manager of Columbia Toy Prod- plaque to Southwestern sales representative Gil Hughes. He chalked up the best gain (42%) of anyone handling the Wonder Horse line in 1963.

BELOW-LEFT: Louis Marx salesman Jerry Fineman and wholesaler Morey Kasch take sides in a battle between a pair of robot boxers. Mr. Kasch's "man" has just been belted.

BELOW-RIGHT: At the Int'l Toy & Trade Fair at the New York Hilton, manning the Fred Bronner exhibit, were Mr. Bronner himself (right); and merchandise manager Charles Resnick. In background are new Matchbox numbers.

In the lower left photo, Marx salesmen battle it out demonstating a Rock 'Em, Sock 'Em Robots set. Playthings, April 1964, p. 58.

SMOKING SPACEMAN: 1950s. 12" high. Battery-operated.

This very desirable, powerful-looking and expensive tin robot was manufactured by Linemar, Marx's Japanese subsidiary, and was also sold under the name of the Japanese manufacturer Yonezawa. Actually, this toy is really a robot rather than a spaceman, but such overlapping is not uncommon in space toys where many spacemen are somewhat robot-like and robots are sometimes called spacemen.

The Smoking Spaceman comes in gray or light blue, with red shoes. As comprehensively described by Lloyd and Glenn Ralston, the toy has rivet detail on his body, red light bulb eyes, a chrome mouth, ears and a combination eyebrow and nose piece with rivet detail. On top of his head, he has a ribbed plastic dome with a rainbow colored cylinder inside it. On his chest he has a chrome coil and a meter [orange and black for the gray version; blue and red for the light blue version]. Below the meter is an on-off switch and he has two chrome gauges on his back. His arms are detailed with many embossed circles. The tabs that hold the arms together have a piece of fold-over metal which hides the joints, making him look much better that the usual robot. His legs house the batteries, one per leg. When he is turned on, the robot walks forward moving his arms in a natural manner, the plastic cylinder and his eyes light up and he stops, he blows smoke out of his mouth three times, and then he resumes walking.[13]

The box for the toy has erratic lettering that is difficult to read. "Spaceman" is spelled out in stylized letters that are not in a straight line, but instead are above and below, as well as overlapping one another. The spaceman is illustrated walking forward in a lunar landscape, while light shines from his eyes and from the cylinder on top of his head.

There is another box for the toy which has a different illustration, although the subject matter is similar. This box has standard lettering and the illustration shows smoke issuing from the robot's mouth that is so dense that part of his right arm is obscured. The wording is the same for both boxes, listing the spaceman's accomplishments: "Lighted Eyes/Stop and Go Action/Lighted Pistons/Smoking." Also on the box are the Linemar name and logo.

Lloyd and Glenn Ralston mention having seen yet a third box for the toy, this one showing a smoking spaceman in the foreground and two in the background. They feel that the use of the words "Lighted Pistons" on the different boxes is not correct because, "A piston moves up and down inside a cylinder."[14] Apparently, the word "cylinder" should have been used instead.

Value: Excellent $1,900-$2,000

Box for the Smoking Spaceman. Notice the erratic lettering. This toy also came in another box with a different illustration and standard lettering. Courtesy LLOYD RALSTON GALLERIES, Fairfield, Ct.

The 1950s Smoking Spacemen measures 12" high and was manufactured by Linemar. When activated, he walks while his arms move, the cylinder on his head and his eyes light up, and he blows smoke out of his mouth. Courtesy LLOYD RALSTON GALLERIES, Fairfield, Connecticut.

SPACEMAN

SPACEMAN: 1950s. 7 1/2" high. Battery-operated.

This tin remote-controlled Spaceman was manufactured by Linemar, Marx's Japanese subsidiary. Lloyd and Glenn Ralston describe his spacesuit as being lithographed in red with dark red bands on his arms and legs. His helmet is white with a red headband and a light built into it. In his [right] hand he carries a blue and silver space ray gun. The gun is jointed in the middle, so that the toy will fit into its small box.

On his left arm, the Spaceman has a wristwatch; he has a flying saucer patch on his shoulder. He has lacings on his legs to hold his red boots to his flying suit. On his right leg is the Linemar logo. Embossed on his chest there are two pouches and a pistol holster attached by a black belt. He has a pair of silver and yellow air tanks with hoses connected to a regulator for his breathing. He walks forward with his arms moving and the light shining constantly. This toy shares the same body, with helmet variations, as a Spaceman made by the Japanese manufacturer TN.[15]

The box illustration for the Spaceman shows him gripping his gun with both hands. Although the light on his helmet is shown, the design of his spacesuit is different, lacking the pouches and showing lines around the waist that the actual toy does not have. As with so many of these toys, the Spaceman stands in a lunar landscape. In the background is what appears to be a space capsule. The box is lettered "Spaceman," "Battery operated remote control and with light on head," and "Linemar Best By Far," and also bears the Linemar logo.

Lloyd and Glenn Ralston have stated above that the Spaceman shares the same body as a spacemen made by the Japanese manufacturer TN. Yet they discovered that, strangely, the illustration of the Linemar box, which differs from the toy itself, resembles closely the illustration for yet another spaceman by TN, this one measuring 9 1/2" high.[16] With the exception of showing the TN Spaceman with an antenna at his back and gripping the gun with only one hand (while the Linemar Spaceman has no antenna and grips his gun with both hands), the illustrations on both boxes are extremely close. Both boxes have the same name for the toy, but TN separates the name into two parts, "Space Man," while Linemar makes "Spaceman" one word.

Value: $400-$500

Linemar manufactured three versions of the Spaceman with wind-up mechanisms. The bodies of the variations are the same as that of the battery-operated Spaceman, but the helmets have larger oval-shaped openings without lights. The belts of these Spacemen are lower on their bodies and each of the Spacemen has a bib-shaped design on his chest.

GRAY SPACEMAN (A): 1950s. 7 1/2" high. Wind-up.

This variation also comes with the markings of the Japanese manufacturer AHI. On the Spaceman's chest, various gauges and meters are lithographed, with a circular one in the middle and two roughly vertical-shaped ones on each side. Around the Spaceman's neck is a decorative design that is continued on the upper part of the opening on his helmet. None of the other Spacemen bear such a design. Extending from the left side of his helmet is a spring that probably leads to an oxygen tank on his back. A spring also extends from under his neck to above his belt.

Value: $600-$700

BLUE SPACEMAN (B): 1950s. 7 1/2" high. Wind-up.

This Spaceman has more detail on his body than the others. Even the prongs of his belt buckle can be seen. Attached to his belt is a lithographed coil of rope. Above his belt are what look like various tools; a wrench can be distinguished. On his chest is a bibbed area with a somewhat striped design and a single dial, looking rather like a golf ball on a tee.

Value: $600-$700

This 1950s battery-operated Spaceman was manufactured by Linemar and measures 7 1/2" high. He walks forward while his arms move and the light on his helmet shines. Courtesy LLOYD RALSTON GALLERIES, Fairfield, Connecticut.

GREEN SPACEMAN (C): 1950s. 7 1/2" high. Wind-up.

This Linemar Spaceman was also marketed under the name of the Japanese manufacturer Naito Shoten, as shown in the book *YESTERDAY'S TOYS, ROBOTS, SPACESHIPS AND MONSTERS* by Teruhisa Kitahara.[17] The well-known auction firm of SMITH HOUSE TOYS, Eliot, ME, lists the toy as being made by the Japanese manufacturer AN.

This toy is the most interesting of all the Spacemen, for unlike the others he does not have dials and meters on his chest, but instead has stylized bullets. He also has a spring projecting right out of his chest to his stomach and another from his helmet to his oxygen tank. Most interesting of all, and perhaps a tad disquieting, is the fact that unlike the other Spacemen with their human features, this one merely has two large predominantly blue circles on a black background instead of facial features. These blue circles are glimpsed through a red-edged opening in an orange helmet. Around the helmet's neck is a lithographed rivet design. The gun that the Spaceman holds in his blue-gloved right hand is orange and red. His green spacesuit has orange, white, blue and black trim. The Spaceman's oxygen tank and boots are red. An altogether intriguing toy!

Value: $600-$700

This 1950s Linemar wind-up walking Spaceman measures 7 1/2" high. Springs protrude from his body and instead of facial features he has two blue circles. Courtesy SMITH HOUSE TOYS, Eliot, Maine.

OTHER SPACE TOYS

ASTRO BUGGY: 1968. 12" long. non-powered.

The plastic Astro Buggy was sold as part of a set in the 1968 *Montgomery Ward Catalog* (p.277) along with the Moon Crawler (see below) and the Jane and Johnny Apollo Astronauts (see Robots and Other Space Figures). However, each of these toys could be purchased separately. The ad described the toy as follows: "Big vehicle features full-view, flip-down green tinted canopy, contour seat, fully instrumented panel and flotation-type balloon tires. Perfect for moon travel." The toy sold for $2.99. Interestingly, this toy did not appear in the *Sears Catalog* at the time.

The most noticeable aspect of the Astro Buggy is the exceptionally large plastic balloon tires (the exact size is unknown). These have embossed treads. On the front of the vehicle is the number "6." In the ad, a doll is shown seated inside the vehicle, but the doll did not come with the toy.

Value: $125-$175

EARTH SATELLITE PROTOTYPE: 1958. 7" long, 6" wide, 4" high. Synchron (electrical) motor.

This interesting prototype has a barn-like shape. It is hand-painted to resemble a starry blue sky. The only writing on the toy is "Earth Satellite" in cream and red letters. The satellite is shown orbiting a small section of the earth. A hand-written paper tag attached to the prototype's cord reads, "X-5288-D/Erie/5/16/58."

The underside of the prototype houses the synchron motor. Of particular interest is that it can be clearly seen that this item was made from the bed of a stake truck.

Just what Marx had in mind for the prototype is not known. The use of the motor might perhaps indicate that the item was planned eventually to be a store display.

Value: ?

Earth Satellite Prototype was made in 1958 and measures 7" long, 6" wide and 4" high. It is run by a synchron motor. R. Johnson collection, Robert Grubb photo.

Underside of the Earth Satellite Prototype showing the stake truck from which the item was made. R. Johnson collection, R. Grubb photo.

MOON CRAWLER: 1968. 13" long. Battery operated.

Like the Astro Buggy above, the Moon Crawler also came as part of a set along with the figures Jane and Johnny Apollo. Also like these toys, the Moon Crawler could be purchased separately.

The 1968 *Montgomery Ward Christmas Catalog* (p.277) advertised the toy for $4.99. The ad described the Moon Crawler as, "Designed for rough terrain. 10 rotating legs overcome all barriers; steers right or left. Rugged plastic with full-view flip-open green canopy."

The Moon Crawler is an interesting looking vehicle. The canopy is shaped like a snowdome. Underneath are the markings, "U.S. Space Corps." At the side are the rotating legs with what appear to be wheels at their centers. At the bottom of each leg is a treaded square to help grip the ground.

Value: $150-$225 when marked NASA

MOON PATROL HELICOPTER: 1968. 10 1/8" long. Wind-up.

This tin toy is predominantly white with red, blue and silver trim. "Moon Scout" is shown on the top sides and the NASA insignia is on the rudder. Astronauts are lithographed in the cockpits and windows. The toy has yellow front and rear rotors which spin as the helicopter rolls along on black rubber wheels. The Marx-Japan logo is on the bottom of the toy.

The box for the toy has an illustrated lid, while the rest is plain. The lid shows the helicopter flying above the moon. Marx enthusiast John Ritter claims that such a feat would be physically impossible because the moon has no atmosphere. The box is marked with two different sets of numbers: "48-23279" and "J-6411."

Value: Excellent with box: $100-$200

The 1960s wind-up tin Moon Patrol Helicopter is 10 1/8" long. Rotors spin as toy rolls along. R. Johnson collection, R. Grubb photo.

MOON SCOUT HELICOPTER: 1950s-1960s. 7" long. Battery operated.

This remote controlled predominantly tin toy has a revolving rotor. Like the wind-up Moon Patrol Helicopter, its body is marked with the words "Moon Scout" and the NASA insignia. Although similarly marked, the Moon Scout Helicopter does not resemble the Moon Patrol Helicopter, since it has a somewhat square body compared to the Moon Patrol Helicopter's longer body. The action of the toy is to go forwards and backwards.

The illustrated box for the toy shows it flying over a lunar surface. As previously stated for the Moon Patrol Helicopter, such an action is physically impossible because the moon has no atmosphere. Crew members shown in the windows of the vehicle are not on the actual toy. Wording on the box is "Moon Scout Helicopter/With Engine Noise and Revolving Propeller/Battery Operated Remote Control."

Value: Excellent with box: $60-$75

MOON SCOUT TRUCK SET: 1968. About 8" long each. Battery operated.

Both white plastic trucks are marked "Moon Scout" on the front. In between the two words is the United States flag. The trucks have large clear domes for good viewing. The sides of the trucks are marked "NASA." The trucks are able to move forward or backward.

The trucks were advertised in the 1968 Montgomery Ward Christmas Catalog priced at $6.49 a pair. They were described as having deeply treaded [plastic] tires that could climb easily over obstacles. One truck has two movable arms for probing the planet's surface, while the other has a silver colored scanner on the back that can be removed. Four astronauts measuring 3" high each came with the set. An additional two astronauts drive each truck. The ad does not mention the astronaut drivers, although they are illustrated. The truck with the scanner has been seen in a box marked "Moon Grabber."

Value: Set with box: $50-$75+

MYSTERY SPACE SHIP: 1962. 7 1/2" diameter. Gyro powered.

This interesting plastic toy is a yellow gyroscope, designed to resemble a flying saucer. The toy comes with a green base embossed to appear like a lunar surface, and also a stand and holder referred to as a gimbal. However, the toy can perform without these extra items.

The Mystery Space Ship comes with many additional green and orange accessories which are conveniently listed on the instruction sheet, including two space cars, a rocket spring, a rocket launcher and three rockets. The rocket launcher can be mounted on either the gyroscope or on the base. In addition, the toy comes with nine Moon Men and nine Spacemen, each approximately 1 1/2" high (one particularly interesting man has four arms), plus two orange 1" Astronauts who are sealed inside two movable plastic bubbles on the space ship.

This toy is unique. Writing in *Antique Toy World*, Tom Frey points out why: "With the design of a flying saucer, the Mystery Space Ship has the distinction of being the first known space toy to use a gyro motor. To operate it, a hand crank is inserted in the ship's underside. As it is gradually turned, the pitch of the whirring motor intensifies as does the force of the gyroscopic power." The toy can actually perform fifty-two tricks as shown in the illustrated blue paper instruction sheet which measures an impressive 17" long and 22" wide. Tom Frey states that the instruction sheet gives scientific explanations of centrifigal force, gyroscopic principles and torque vectors, and then quizzes you on the information. If all three tests are answered correctly, you earn, respectively, co-pilot's, pilot's and interplanetary navigator's licences.[18] Examples of the tricks the toy can perform are balancing on a stout string, rolling in a circle if rolled at an angle, and spinning like a top.

The 1962 *Sears Christmas Catalog* advertised the Mystery Space Ship for the rather uneven price of $3.66. Additional information about the toy provided by the ad was that the gyroscope spins at 3,000 revolutions per minute.

The box for the toy is yellow, orange, white, green and black. In various color blocks on the sides of the box, there are comments such as, "Learn About 'Precession' (sic) – It's Fascinating" and "Powerful Rotor Will Support Three Lbs." Differ-

Manufactured in 1962, the 7 1/2" diameter Mystery Space Ship is gyro-powered. The toy can perform 52 tricks. Samuel Caley collection, R. Grubb photo.

ent parts of the toy are shown, for example, the Rocket Launcher, Moon Men and Astronauts. Wording is, "Your Own Moon Base/ Gyro-Powered – It Defies Gravity."

The front of the box has the toy's name "Mystery Space Ship" by Marx, plus the Marx logo, name and address. Noteworthy is the phrase, "As Seen On TV." Considered a plus, it is an interesting contrast to the company's former resistance to using media advertising.

The Mystery Space Ship is illustrated on the box with a starry black sky in the background. The figures shown are all green, although some of the real figures are orange.

Value: Excellent with box: $160-$200

PROJECT MERCURY RECOVERY SET: 1960s. 5-8" long. Non-powered.

This plastic set consists of a 5" long silver space capsule and an 8" long helicopter. The bottom of the capsule is removable so a figure can be placed there if desired. The set comes on a 17" long by 11" wide blister pack. *Hake's Americana & Collectibles Catalog No. 111* for November 1990 states that this Marx set was modeled after the first U.S. manned space project of 1961.

Value: Mint: $50-$100

REX MAR'S SPACE TARGET GAME: c. 1950s. 13 1/2" long target. 11 7/8" long gun.

This toy is described by Rick Johnson as being made by Marx in tin and plastic. The rifle is blue plastic and shoots wood bullets such as those used in the prewar anti-aircraft gun.

The target is lithographed tin, depicting a space city. It is marked "Rex Mars Planet Patrol." Point values are indicated below 5 aliens who are suspended on a rod across the top. Each alien is a different shape and color and spins head over heels when hit. The plastic aliens are 3 1/2" tall.

The box is plain with blue lettering and Marx logos.

Value: $200-$300

This tin and plastic 13 1/2" long target game was manufactured in the 1950s. R. Johnson Collection. R. Grubb Photograph.

ROCKET EXPRESS: 1950s. Approximately 5" square. Mechanical.

The tin Rocket Express made by Linemar is similar to the Marx Honeymoon Express toys except it does not consist of a train that goes through tunnels while an airplane circles above. Instead, two rockets speed through two tunnels while a satellite, instead of a plane, flies overhead. Another difference is that the Rocket Express is smaller and has a 5" (approximately) square base instead of the 9 1/2" diameter circular base of the Honeymoon Express.

The Rocket Express is basically pink and blue and is lithographed with various rockets and planets in colors of red, white and yellow.

The box for the toy reads, "Rocket Express/Linemar/ Best By Far." Illustrated is a lunar landscape with a space station, satellites, rockets and planets.

Value: $700-$1000

ROCKET PROTOTYPE: 1950s ? 15 1/4" long. Friction.

This prototype plastic rocket, as described by Rick Johnson, is red and silver with a green rubber nose spear. The Prototype has a metal motor casing and the engine unit is visible through the transparent hatch on top of the fuselage and can be seen rotating as the rocket rolls along on black rubber wheels. In the cockpit is the yellow plastic head of the pilot. Handwritten markings underneath are "X-2888 Erie (?) PA."

The vehicle appears to be adapted from a plastic airplane Marx produced in 1952 called Sparking Supersonic Jet Plane. That plane also works by friction and also has a transparent section behind the pilot which shows the revolving motor.

However, although the bodies of the two toys are similar, the Rocket Prototype lacks the plane's wings, retaining only the pods which are close to the body. In the plane, the pods are at the ends of the wings. Another difference between the two toys is the shapes of the stabilizers, with the Prototype Rocket having a more angularly-shaped and larger stabilizer than that of the plane.

The 1950s Rocket Prototype measures 15 1/4" long and has a friction motor. The engine unit is visible as the rocket rolls along. R. Johnson collection, R. Grubb photo.

SATELLITE INTERCEPTER PROTOTYPE: 1960s. Base 4 1/2" x 5," Blower 5 1/4" high, Launcher 6" long. All tin. Battery operated.

This prototype was made by Linemar. The launcher is hand-painted metallic blue with black details, while the blower is hand-painted light grey with mostly black details. The blower base (battery box) is hand-painted dark blue and depicts spaceships, planets, stars and the moon's surface. The artwork bears a certain similarity to Lindberg Model Kits.

To operate, the switch in the base of the launcher unit turns on a blower in the other unit. (The two units are connected by a two-conductor wire.) The blower rotates above its base, which contains the battery box. The blower makes the satellite float above it. The objective is to look through the cross hairs in the launcher and try to shoot down the orbiting satellite (a styrofoam ball) with rocket-shaped darts.

Rick Johnson believes that the Linemar prototype, which he described above, is likely derived from a toy "Satellite In Orbit" marked "SH" produced in Japan. This toy has an on/off lever slot in the base which is evident in the Linemar prototype, but not in the production Satellite Intercepter toy. There are other similarities with the blowers and bases. The bases are actually the same size.

"Satellite In Orbit" toy, produced in Japan, is marked "SH." It has the same size base and a blower similar to the Linemar prototype. R. Johnson collection, R. Grubb photo.

The battery operated 1960s tin Satellite Intercepter Prototype has a 5 1/4" high blower and a 6" long launcher. The objective is to look through the crosshairs in the launcher (right) and try to shoot down the orbiting satellite (a styrofoam ball) with darts. R. Johnson collection, Robert Grubb photo.

SATELLITE INTERCEPTER: 1960s. Base 4 1/2" x 5," Blower 5 1/4" high, Launcher 6" long. All tin. Battery operated.

The toy was produced by Linemar and works in the same manner as the Satellite Prototype described above. A switch in the base of the launcher unit turns on a blower in the other unit. The two units are connected by a two-conductor wire measuring 36" long. The blower rotates above the base which contains the battery box. The blower shoots out air which makes the satellite, a styrofoam ball, float above it. The objective is to look through the crosshairs in the launcher and try to shoot down the orbiting satellite (styrofoam ball) with two red plastic rocket-shaped darts.

Rick Johnson explains how the Satellite Intercepter production toy is subtly revised in shape from its prototype, as follows:

> lateral bracing is added to the launcher supports; there is a wire hole in the base in a different location; the bigger housing is slimmer and the corners more rounded; the blower is reconfigured to give the satellite a wider orbit; the lithography is entirely different from the prototype paint scheme.

The launcher is silver with multicolored details including depictions of astronauts, missiles and the globe. A knob on the side of the launcher swivels it up and down and the base can swivel.

The blower has windows and is predominantly beige and red. The blower base (battery box) is nicely lithographed with depictions of a moonscape in orange. Within this terrain is the Linemar logo. In the background are silver and dark blue mountains. Above this is the blue sky filled with heavenly bodies (stars). Also shown are astronauts, rockets and a space station, all illustrated in a different style than on the prototype.

Value: Excellent with box: $250-$350

SATELLITE LAUNCHING CAR: 1950s-1960s. 5" long. Friction ?.

Made by Linemar, the tin Satellite Launching Car is somewhat truck-like in appearance. It is predominantly silver with red, white and black trim. The blue plastic satellite is mounted to the top of the car. When the car is pushed, the satellite is launched.

Value: $25-$50

SP-1 SPACE CAR: 1950s. 6" Long. Friction.

The attractive SP-1 Space Car is predominantly blue with red and black trim. Portholes are lithographed on the sides as well as "SP-1." In profile, the vehicle has a somewhat tank-like appearance with fins on each end which are reminiscent of the 1950s cars. Mounted on the front of the vehicle is a spark-shooting gun. As the SP-1 Space Car moves, its engine makes a noise.

The illustrated box for the toy shows the SP-1 Space Car flying over the lunar surface with gun firing.

Value: Excellent/Near Mint with box: $700-$800

SPACE MISSILE: 1950s-1960s. Approximately 1 1/2" long. Non-powered.

This tiny tin toy was manufactured by Linemar and belongs to a series called Elegant Miniatures. The wheeled missile is predominantly red with white trim.

Markings on the box are "Elegant Miniatures No. 111 Space Missile," the Linemar name and their logo.

Value: Excellent with box: $20-$30

Battery operated 1960s Satellite Intercepter has two units: a 5 1/4" high blower which shoots out air, causing a satellite (a styrofoam ball) to float above it, and a 6" long launcher to shoot darts at the satellite. R. Johnson collection, R. Grubb photo.

SPACE RANGER: 1962. 6 foot orbit. Battery operated.

The Space Ranger toy puts one in mind of the Marx Tower Flyers manufactured through the 1920s to the 1940s. Similar to these earlier toys, the Space Ranger has a tower-like structure with a pivoting bar. A plastic rocket is at one end of the bar and a counterweight at the other. Unlike the earlier toys, the Space Ranger has an embossed base resembling lunar terrain, and a platform near the top of the tower. Above the platform are the markings "X-21." This toy was advertised in the 1962 *Sears Christmas Catalog* (p.451) and is described somewhat sparsely in comparison to many of their ads. For instance, only one measurement is given. However, the Marx name is shown, which is not a frequent occurence. The ad describes the Space Ranger in the following way: "Send this plastic rocket ship into 6-foot orbit. You're at the remote controls as metal bar swings it forward, into reverse. It blows up a flying saucer, rescues a man, sends a secret message. Accessories - exploding flying saucer, nose cone, space car, 3 rockets, 6 astronauts, 6 moonmen, etc." The toy sold for $10.99, a fairly high price for the time. Six other space toys advertised on the same page are priced below this toy.

One of the accessories in the ad, it is not clear which, appears to be similar to the base of the Mystery Space Ship. The figures sold with the toy are the same as those belonging to the Mystery Space Ship. Such similarities would not be strange since the Mystery Space Ship was sold at the same time as the Space Ranger. Some of the figures have loops at their tip which hook onto the rocket, so they can be "rescued."

The control box is connected through a cable into the tower base. The rocket ship can be made to stop, hover, reverse, go fast or slow or go into orbit. In addition, the rocket ship can drop bombs by remote control on the "enemy" flying saucer on the ground. When hit, the saucer flies apart by mechanical spring action. In turn, manually operated rockets from the ground can be fired at the rocket ship. The scout figures can be lowered to a space car by a manual winch. The remote controlled rocket nose cone can become a warhead missile, detaching from the rocket and becoming an "explosive." It is also possible to lower the rocket to the ground by remote control.

Value: $500-$600

SPACE RANGER EXPLODING FLYING SAUCER ACCESSORY PROTOTYPE: c. 1962. 5" diameter. Exploding mechanism.

The exploding mechanism is inside the cream colored flying saucer. Also inside are finely detailed drawings, in pencil, of the interior.

For a photograph, see the Space Ranger Rocket Accessory Prototype below.

SPACE RANGER ROCKET ACCESSORY PROTOTYPE: c. 1962. 7" long. Non-powered.

The white rocket has a hand-made canopy and the front is handpainted in gold with a bluish-green rubber nose spear. Portholes on the body are hand-inked. Also on the body are several stars and bars decals and a label reading "U.S. Airforce X6898." On the production toy, this rocket came with the addition of a circular attachment which fit over it at the base.

SPACE SATELLITES WITH LAUNCHING SITE AND ROCKET: 1950s. 12" long x 9" wide. Crank action.

The tin launching site and base are colored red, yellow, white, blue and black. The toy comes with three 4 1/2" plastic satellites. When one of these is placed on a cylinder on the launching site, and a crank on the side of the site is turned, the satellite flies into the air for a few seconds and a siren noise is sounded. Also in the base at one end is a white rocket and at the other end of the base is a grey radar screen.

The base is nicely lithographed with scenes of a satellite launching pad. Among the markings are "Service Ramp" and "Caution" plus "7" and "8." The launching station is marked with the number "2" and is lithographed with windows and personnel, some looking out of the windows, while others walk by. The pad is lithographed with trucks, fire engines and satellites. In the service ramp section, rockets are being serviced.

The launching site building is rectangular with an archway. Attached to it is another part of the building, rather wall-like in shape. Markings are "Launching Station" and "Space Satellite" plus the number "2."

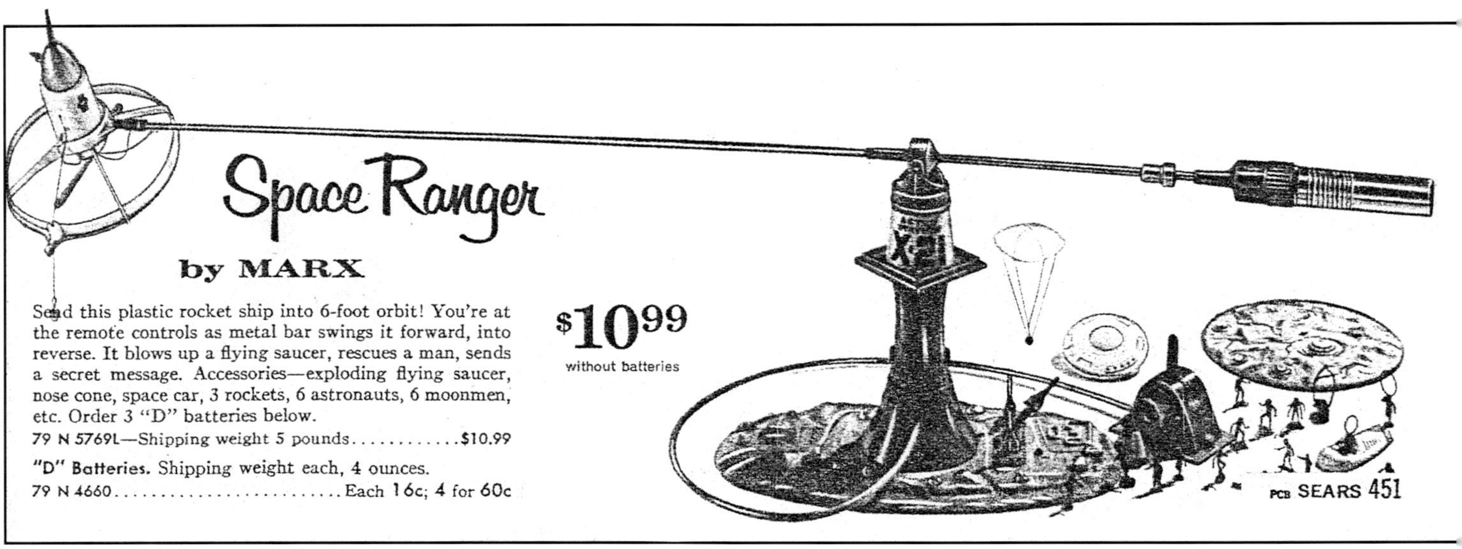

Space Ranger advertisement from the 1962 Sears Christmas Catalog (p.451).

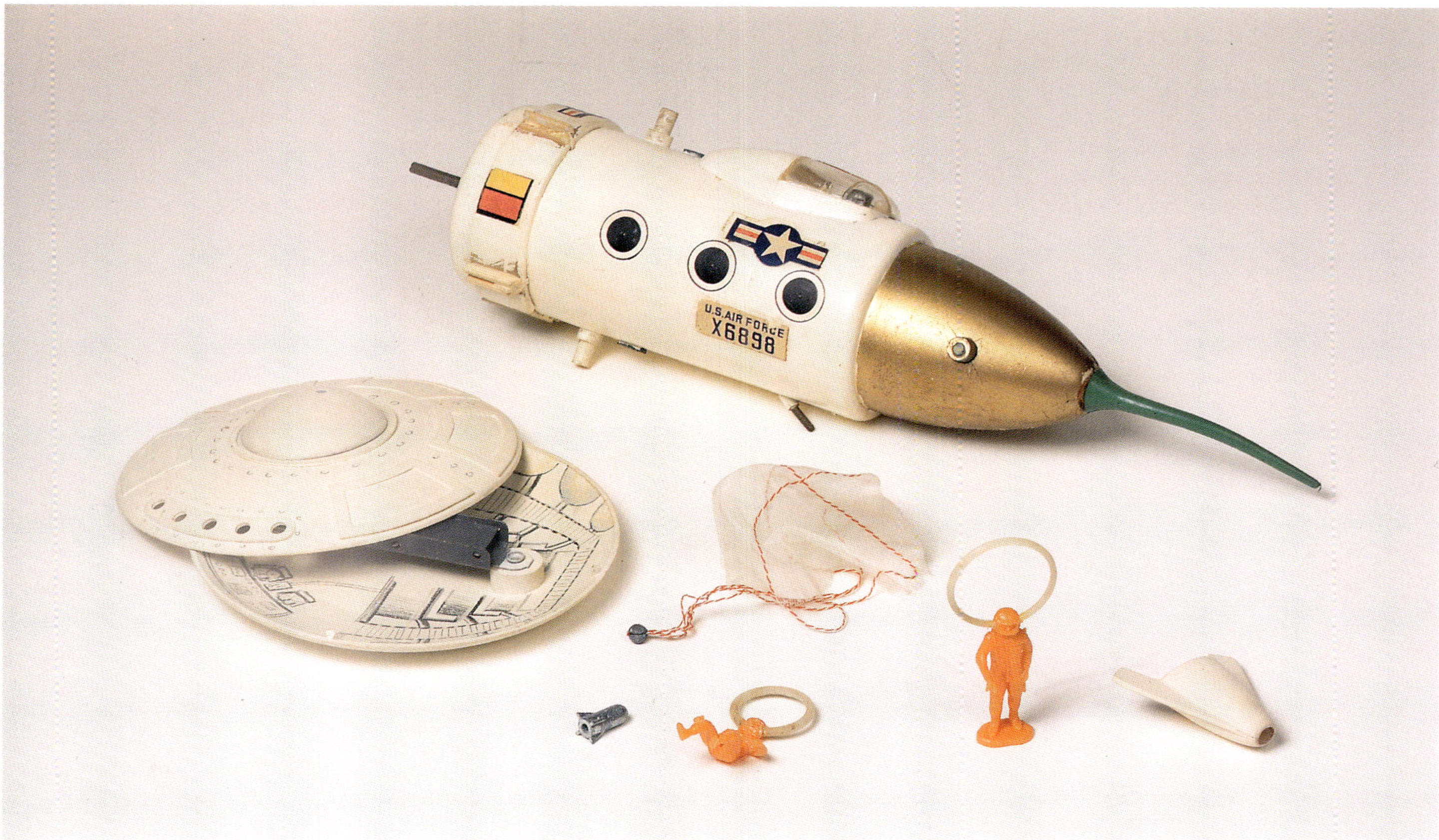

Space Ranger Rocket Accessory Prototype, 1960s (on left), measures 7" long and 1960s Space Ranger Exploding Flying Saucer Accessory Prototype (on right) has a 5" diameter. The other items are from the production Space Ranger toy and consist of figures, a parachute, an assembly to attach a counterweight and, possibly, a nose cone. R. Johnson collection, R. Grubb photo.

The same building and base, but with different lithography, was used in a 1954 toy which featured helicopters instead of satellites.

The box for the toy reads "Space Satellites With Launching Site" and shows illustrations of rockets and satellites flying over a city.

Value: $400-$500

SPACE SATELLITE WITH LAUNCHING STATION: 1956. 12" long x 9" wide. Crank action.

This toy works with the same crank action and has the very same tin building and base as used for the "Space Satellites with Launching Site and Rocket" described directly above. The difference is that the 4 1/2" diameter satellites have been replaced by one, or possibly more, 11" red plastic satellites.

Sears introduced the toy in their 1956 *Sears Christmas Catalog* (p.268), where it was advertised to sell for $2.89. The ad makes it clear that these satellites have a different appearance as they have "transparent plastic domes to show their mysterious inner workings; safe soft plastic rims." The rocket and radar screen are not mentioned or shown in the ad, but instead there are five plastic spacemen whose size is not indicated. It is possible that this toy was also made in a version without the spacemen.

Another addition is that the toy has a metal sign at the end of the base which gives notice to all to stay clear of the spacecraft at takeoffs and landings.

The box for the toy reads "Space Satellite With Launching Station" and shows the satellite in the sky and the launching station below.

Value: ?

SPACE SHIPS: 1950s. Approximately 5" long each. Non-powered.

Futuristic, exotically shaped and painted in metallic colors, these four plastic Space Ships are quite engaging. They come in varying shapes and are colored variously maroon, gold, silver and green. They are "bin" toys, that is, toys that were sold individually in stores such as Woolworths. The plastic wheels and Marx logo are on the undersides of the toys.

Marx manufactured four Space Cars which are the same size and have the same metallic coloring as the Space Ships.[19] Marx also manufactured airplanes which are close in size and which are painted in metallic colors, but except for one plane having the same silver color, the others have a different blue color.

MAROON SPACE SHIP:

The maroon Space Ship is particularly interesting in that it is the only one of the four to have twin fuselages. The area of the wing between the twin fuselages also has a multiple row of windows. Each fuselage has multiple window rows with the fuselage cabins emerging from a bat-shaped wing. The wing tips (pointing to the rear) come to sharp points. There is a single high rudder and two motor jets are on the rear edge of the wing.

GOLD SPACE SHIP:

There are several differences between the gold and maroon Space Ships. To begin with, although the Gold Space Ship has only one fuselage instead of two like the Maroon Space Ship, the shapes of the fuselages in both ships are similar. However, the windows on the Gold Space Ship resemble portholes rather than the smaller rows of windows on the Maroon Space Ship.

Next, the gold toy has a modified bat-shaped wing with the wing tips pointed forward rather than to the rear as on the maroon toy. Then, there are twin rudders on the gold ship which are not on the maroon one. Also, the twin rudders are not as high as the single rudder of the maroon ship. A further difference is that in the gold model, the engines extend across the full width of the wings rather than the shorter jets extending from the trailing edge of the maroon model's wing. Finally, extending from the forward part of the motor housings of the gold ship are what appear to be weapons.

SILVER SPACE SHIP:

The Silver Space Ship is the strangest looking of all the space ships in this series. It is rather chunky in appearance and in profile it somewhat resembles a tropical fish with a bloated belly, or perhaps a prehistoric reptile. The author has never seen any ship like it. One of the airplanes mentioned earlier, the silver model, closely resembles this toy in appearance.

The fuselage is embossed in a cinderblock pattern. Each side of the fuselage has four windows plus a door with a window in it. Each door opens onto a parabola-shaped (or as Rick Johnson calls it) reverse-delta wing with a ladder leading from the door to the wing tip. There is a large extended canopy on the top forward section of the fuselage. The canopy has many windows. On the forward part of each wing where it meets the fuselage, is a small filler. On the trailing edge of each wing is also a filler. There are three fin-like protuberances (rudders, perhaps) which extend from the top of the fuselage, adding to the ship's fish-like appearance.

GREEN SPACE SHIP:

The Green Space Ship is the most realistic looking version of all the space ships in this series. It has triangular-shaped wings with turned-up wing tips. The fuselage sides are fluted and there is a large square-shaped rudder.

Value: $10-$20 each for ships, cars and planes.

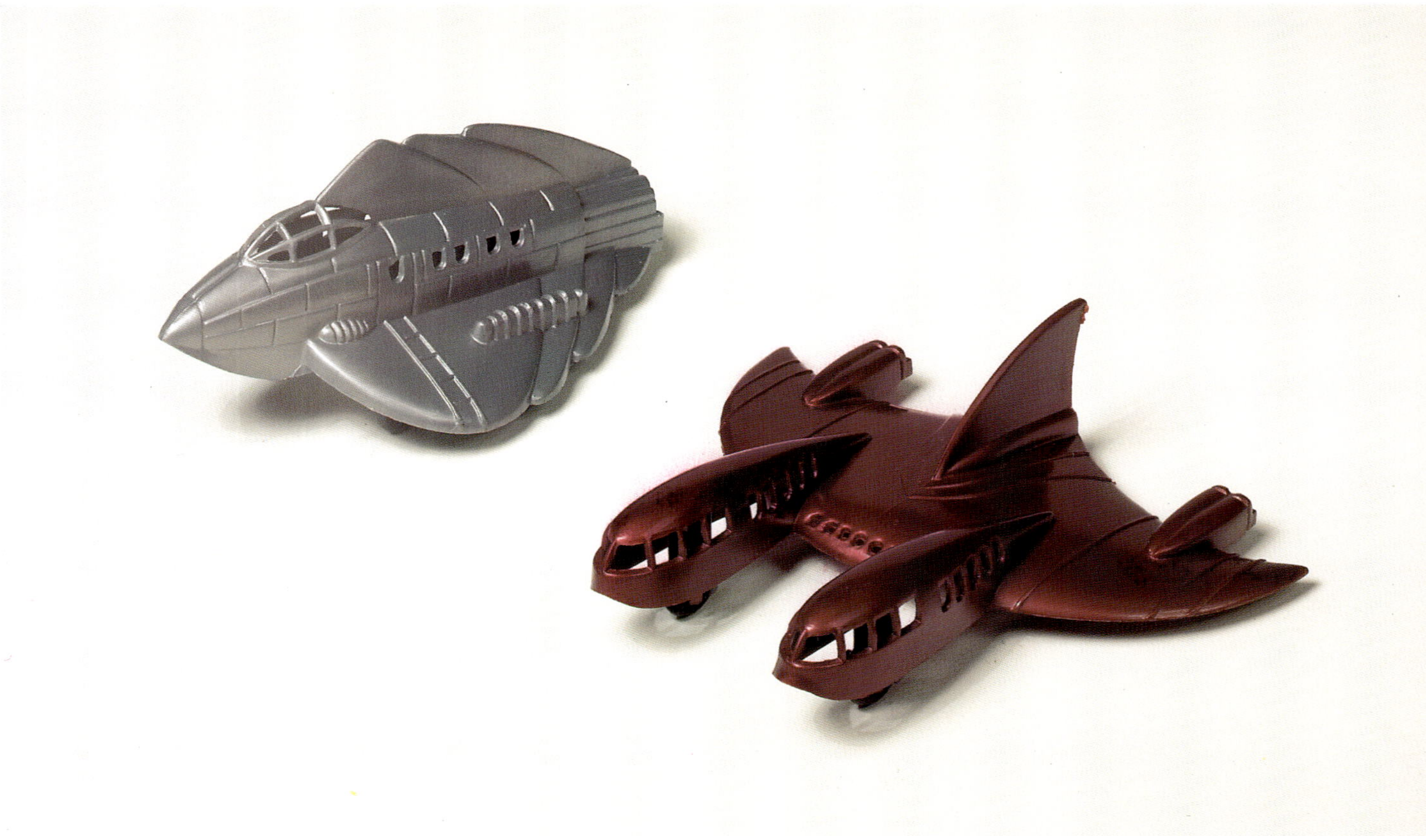

Plastic wheeled Space Ships (1950s) measure approximately 5" long. In addition to the silver and maroon ships shown in the photograph, there are also gold and green ships. R. Johnson collection, R. Grubb photo.

Made in the 1950s, these 5" Marx plastic futuristic cars are a complete set. They are the same size and have the same metallic colors as the space ships described above. The set consists of a coupe, sedan and two convertibles. Trip Riley collection.

Four Marx airplanes are close in size and painted in metallic colors like the space ships. However, only one plane has the same silver color. Others are blue. The complete set consists of a B-47 Stratojet in silver, 5 3/8" long; F-86 Sabre in blue, 5 1/2" long; F-89 Scorpion in blue, 4 5/8" long; F-90 Penetration in blue, 6 7/8" long. R. Johnson collection, R. Grubb photo.

TOM CORBETT POLARIS SPACE SHIP: 1952. 12" long. Mechanical.

The tin Tom Corbett Space Ship is similar in size and shape to the earlier Buck Rogers Ships of the 1930s. Unlike those ships, however, the large side wings have been discontinued, so the blue tin wheels can be clearly seen.

In the 1952 *Sears Catalog* (p.258), the toy sold for $1.69. The ad described it as:

> "An authentic reproduction of the 'Polaris' rocket space ship of Tom Corbett, popular TV personality...all-over lithographed design shows Tom Corbett at the controls with other space cadets. Cadet insignias on rocket. Sparks shoot from tail exhaust when space ship shoots across floor... raised dummy observation domes. Two large side wheels, one small rear wheel."

The toy is primarily blue, yellow and red. On the top there are four red bubbles and the Marx logo is on the nose of the ship. The number "2" is repeated on the top and sides. Wording on the sides in yellow is "Tom Corbett Space Cadet" with a small rocket passing over the "C" of Cadet. Near the end of the ship is "Polaris." Stars twinkle in the blue background. There are small fins, or wings, at the front and back sides of the ship.

The box for the toy shows the uniformed Tom Corbett and two fellow cadets, their hands gripping their belts. The globe is in the background. Wording is "Tom Corbett Sparkling Space Ship" with a star dotting the "i" of sparkling.

Value: Excellent with box: $475-$800

The early 1950s tin windup Tom Corbett Polaris Space Ship is 12" long. It is similar to the Marx Buck Rogers ships of the 1930s. Dr. Malcolm Kates collection, R. Grubb photo.

TWO-STAGE EARTH SATELLITE: 1957. Rocket 8 1/8" long, Base 9 1/4" long. Friction.

This interesting tin toy was made by Linemar. As described by Rick Johnson, the rocket is red with white and blue fins and black rubber wheels. The Linemar logo is under the left fin. The toy is marked on the side in white "Two-Stage Earth Satellite."

The launching station, which is marked as such, is yellow with a crank on the back. One part of it has a lithographed brick-like pattern and steps. On the back, near the steps, is another Linemar logo. The other part of the launcher has a lithographed rivet design. The launching station is attached to a green and grey base. To operate, the tail of the rocket is snapped against the launching station, the crank is wound clockwise and the release button is pushed, then the rocket starts running, while a siren sounds. After running three to four feet, the rocket separates automatically into two pieces.

The box lid for the toy is red, yellow and blue. Wording is "Two-Stage Earth Satellite With Launching Station." Also on the box is "Linemar Best By Far." The satellite is illustrated in its separated state, with its name in blue letters instead of white as on the actual toy.

Value: With box: $500-$700

Friction Two-Stage Earth Satellite, made by Linemar in 1957. Satellite measures 8 1/8" long. When activated, satellite moves forward and separates into two pieces. R. Johnson collection, R. Grubb photo.

TWO-STAGE LAUNCHING PAD: 1950s. 7" long base. Battery operated.

This tin toy by Linemar (also marketed by the TN – a Japanese manufacturer) is unusual in that it features a three-dimensional man seated in a chair before a screen. The man wears a blue suit and the chair on which he sits is yellow. On his right are two plastic and rubber red and yellow rockets and a launcher marked "Jupiter." Underneath is a grey base nicely lithographed with a scene on a screen of the rocket heading for outer space. Underneath is the word "Jupiter." Another lithographed scene shows two workers, one with earphones, surrounded by various dials and meters.

To operate, the worker seated on top of the base pushes a lever with his movable left arm. The screen in front of him lights and a rocket is shot out of the launcher.

The box for the toy is illustrated with two workers eagerly watching on a screen the launch of the rocket. The name "Two-Stage Rocket Launching Pad" is in red, outlined with white.

Value: With box: $400-$700

The 1950s battery operated Linemar Two Stage Launching Pad measures 7" long. To operate, the seated figure pushes a lever, the screen lights up and a rocket shoots out of the launcher. Courtesy SMITH HOUSE, Eliot, Maine.

WHISTLING TOP SET: 1950s-1960s. Tops are 1 7/8" diameter each, Activator is 4" long. Friction.

In the 1930s, Marx manufactured a top set, one of which had a lithographed face. In this 1950s-1960s tin top set, manufactured by Linemar, both tops have faces, really just eyes and circular lines which pass as mouths.

The spherical tops are marked "Freedom" and "Liberty Bell" after the Mercury Space Capsules of those names. Rick Johnson describes the "Freedom" top as orange with blue and yellow stripes. Above the eyes is the marking "A-1802." The "Liberty Bell" top is blue with red, dark blue and yellow stripes. Above the eyes, in much larger markings than that of the other top, is "S28." The Linemar logo is on the bottoms of the toys. The activator is black with an unpainted metal plunger.

To operate, the shaft of the top is placed into the activator. The plunger is squeezed, thus spinning the top. The spinning top drops out of the activator onto a surface and continues spinning.

The box lid for the tops is dark blue and illustrates the tops in orbit around the earth. In the background are heavenly bodies. Wording states that the contents consist of two automatic Whistling Tops with Activator.

Value: Excellent with box: $150-$175.

Linemar tin friction Whistling Space Top Set. Tops are 1 7/8" diameter each and are marked "Freedom" and "Liberty Bell" after the Mercury Space Capsules. R. Johnson collection, R. Grubb photo.

FOOTNOTES:

1. "The Robot Report and Astronaut Account" by Lloyd and Glenn Ralston, *ANTIQUE TOY WORLD,* February-March 1981, Vol. 11, No. 2, p. 28.
2. Teruhisa Kitahara, *YESTERDAY'S TOYS/ROBOTS, SPACESHIPS AND MONSTERS,* (San Francisco: Chronicle Books, 1989), p. 7.
3. "Robots" by Donna Kaonis, *COLLECTORS' SHOWCASE,* March-April 1985, Vol. 4, No. 4, p. 38.
4. Ibid.
5. *PLAYTHINGS* magazine, November 1963, p. 85.
6. An eccentric car the "Old Jalopy" had the names of Marx personnel all over it while a roadster was called "Cannon-ball Keller" after Bill Keller, the general manager of the Marx Erie factory.
7. "The Robot Report and Astronaut Account" by Lloyd and Glenn Ralston, *ANTIQUE TOY WORLD,* March 1980, Vol. 10, No. 3, p. 30.
8. "The Robot Report and Astronaut Account" by Lloyd and Glenn Ralston, *ANTIQUE TOY WORLD,* February-March 1982, Vol. 12, No. 2, p. 62.
9. "The Robot Report and Astronaut Account" by Lloyd and Glenn Ralston, *ANTIQUE TOY WORLD,* December 1980-January 1981, Vol. 11, No. 1, p.46.
10. Teruhisa Kitahara, *ROBOTS/TIN TOY DREAMS,* (San Francisco: Chronicle Books, 1985), p. 12.
11. "The Robot Report and Astronaut Account" by Lloyd and Glenn Ralston, *ANTIQUE TOY WORLD,* November 1983, Vol. 13, No. 11, pps. 38-39
12. "Those 60's Ideals" by Robert Borowski, *MODEL AND TOY COLLECTOR,* Spring 1990, No Vol., No. 14, pps. 44-45.
13. "The Robot and Astronaut Account" by Lloyd and Glenn Ralston, *ANTIQUE TOY WORLD,* May-June 1981, Vol. 11, No. 5, p. 36.
14. Ibid.
15. "The Robot and Astronaut Account" by Lloyd and Glenn Ralston, *ANTIQUE TOY WORLD,* October 1980, Vol. 10, No. 10, p. 32.
16. "The Robot and Astronaut Account" by Lloyd and Glenn Ralston, *ANTIQUE TOY WORLD,* May 1983, Vol. 13, No. 5, p. 70.
17. Teruhisa Kitahara, *YESTERDAY'S TOYS/ROBOTS, SPACESHIPS AND MONSTERS,* (San Francisco: Chronicle Books, 1989), pps. 18-19.
18. "Classic Plastic Toys" by Tom Frey, *ANTIQUE TOY WORLD,* January 1986, Vol. 16, No. 1, p. 89.
19. Both the Space Ships and the Space Cars can be seen in the June 1991 issue of *PLASTIC FIGURE & PLAYSET COLLECTOR,* p. 10.

Comic, Film & Television Toys

INTRODUCTION:

There is so much overlapping among comic, film and television toys that it is not possible to consider them separately. To take but two examples, "Superman" and "Batman" have been seen in all three media.

Comic, film and television toys are among the highest valued of all Marx toys. Many of the very desirable metal and combination metal and plastic toys were manufactured by Linemar, Marx's Japanese subsidiary. There are numerous outstanding examples of Linemar toys, particularly the Popeye toys. However, it is still possible to buy fairly priced plastic toys. In addition to the toys described in this book, almost all of which are "moving" toys, many Marx "non-moving" plastic figures were made.

ARCHIE CAR: Late 1960s to 1970s. 7" long. Wind-up.

The "Archie" comic strip was originally created by Bob Mentara. In the late 1960s, the Archie Show was showing on television. It continued under various titles until the late 1970s.[1]

The completely tin "Archie Car" was manufactured by Plastimarx, Marx's Mexican subsidiary. The same die had been used in an earlier car, Marx's "Old Jalopy" manufactured in 1950 (see Maxine A. Pinsky, Greenberg's Guide to Marx Toys, Volume II, M.D. Greenberg Publishing Company, 1990, p. 33.) but the lithography of the Archie Car is completely different from that of the earlier car. The Mexican car is cream and sky blue with touches of red and black. Another version has been seen which substitutes dark blue or black for the lighter blue.

Archie's and Veronica's heads appear on the doors on the driver's side. The doors on the passenger's side show Jughead and Betty. Nowhere are these characters' English names mentioned; Spanish names such as Lupe and Lolita are written instead. There is also the name Napolion Solano which probably stands for "Napoleon Solo" a character in "The Man From Uncle." Nowhere is any copyright credit given, the only marking (other than the names mentioned above) is the word "Plastimarx" in tiny letters on the top of the radiator.

Like the "Old Jalopy" car before it, the "Archie Car" is covered with slang phrases. Since the phrases are slang, many of them do not translate easily, for example, "Dochs of gold." However, phrases such as "Adios Micifuz" and "El Immortal" are easily understood.

The driver is the same as that of the "Old Jalopy" and is stylishly dressed in a tan hat, dark blue coat, white shirt and red tie. The black and white tin wheels with lithographed spokes are identical in both cars.

Value: $800-$1000

The completely tin "Archie Car" is a wind-up and measures 7" long. It was manufactured by Plastimarx, Marx's Mexican subsidiary. Samuel Caley collection, Tim Parrish photo.

Another variation of the Archie Car. Courtesy NEW ENGLAND AUCTION GALLERY, W. Peabody, Massachusetts.

BATMAN (see Super Heroes)

BEN HUR SWORD, SCABBARD & SHIELD SET (see Sword, Scabbard & Shield Sets)

CASPER THE FRIENDLY GHOST: 1950s. 4 1/2" high. Wind-up.

Casper the Friendly Ghost debuted in films in 1945 and appeared on television in the 1950s. Reruns of the Casper cartoons can still be seen on television at the present time.

The all tin toy was manufactured by Linemar. Casper is white with a minimum of black lines depicting various features. His hands hold his stomach. The only touch of color appears in his large blue eyes. In action, the toy bounces around as his spring-mounted head shakes.

The box for the toy reads in large letters "Casper" and in small letters "The Friendly Ghost." The box's blue color matches Casper's eyes. Casper is illustrated with his right arm reaching up, and his left arm at his side, a different pose than on the actual toy. The Casper toy is hard to find.

Value: Excellent with box $750-$800

Casper was manufactured by Linemar in the 1950s and is a 4 1/2" high wind-up. Toy bounces around as head shakes. Courtesy Smith House, Eliot, Maine.

CASPER GHOST TANK (TURNOVER): 1950s. 4" long. Wind-up.

Manufactured by Linemar, the pale green tin tank moves quickly in a circular motion, stopping every now and then as the flat tin figure of Casper moves it upwards from the bottom and turns it over.

As described in *HAKE'S AMERICANA & COLLECTIBLES CATALOG #110* for September 1990, the tank has full color depictions of Casper and other Terrytoon or Harvey comic characters. Casper is shown on the turret along with a mouse believed to be Herman and an unknown crow. On the tank's left side is a cat thought to be Katnip and, on the other side, there is one ghost from Casper's "bad ghost trio." Baby Huey is on the front of the tank and Little Audrey is on the rear. The bottom of the tank has hard rubber wheels and bears the copyright symbols of both Linemar and Harvey Comics.

Contrary to the popularly used name, the box for the toy does not read "Casper Turnover Tank" but, instead, reads "Casper Ghost Tank" in red letters. The box is yellow and the front panel shows Casper raising the tank. The side panels show Casper popping out the top of the tank which he does not do in the actual toy, while a scarry ghost looms up behind him.

Linemar made many similar type tanks such as the Superman tank in the 1940s and a later Superman version, plus the Flintstones and Jetsons turnover tanks (see this chapter).

Value of Casper Ghost Tank: Excellent with box $800-$1,200

DICK TRACY TOYS:

Dick Tracy has been in the comics, on television and in films. The "Dick Tracy" comic strip was created by Chester Gould in 1931. "Dick Tracy" was briefly on television in late 1950 and early 1951. Ralph Byrd starred in the television series as he did in the Tracy movies of the 1930s and 1940s. However, these movies never underwent the media blitz of the recent "Dick Tracy" film starring Warren Beatty.

The Marx Company made a number of nice Dick Tracy toys prior to the 1950s and after 1950, the company continued making interesting and attractive Dick Tracy toys.

DICK TRACY CARS:

Most Dick Tracy cars have similar lithography. On the side, Tracy's head is profiled within a badge along with "Squad Car No. 1" and "Police Dept." while "Dick Tracy" is lettered on the hood. Dick Tracy drives, while various other characters from the comic strip are lithographed in the windows. The colors of the car are mostly different shades of blue or green.

DICK TRACY CAR: 1950s. 6 1/2" long. Friction.

This dark green tin car has the usual markings listed above, but is unusual in that it has a plastic top which is yellow in color. The top is similar to that of the circa 1950 Tricky Safety Car (see Volume I). The Tricky Safety Car also comes with a blue or red top, but so far the Dick Tracy Car has only been seen with a yellow top. The front of the car has lithographed lights and the rear has a paper directions label.

Value: $175-$200

Various 6 1/2" long friction Dick Tracy Cars of the 1940s-1950s along with a similar Fire Chief Car. Note the Tracy car with the plastic top, third from the right. Courtesy LLOYD RALSTON GALLERIES, Fairfield, Connecticut.

DICK TRACY POLICE CAR: Early 1950s. 8 1/2" long. Battery operated.

This remote controlled green tin car was manufactured by Linemar. It can go forward and reverse. The car lacks the usual markings, retaining only Dick Tracy's head within the badge on the car's sides. On the top, in a plastic housing, is a flashing spotlight. The car makes a siren noise.

The box illustration shows the car with a background of skyscrapers. The Linemar slogan, "Linemar Best By Far" is also on the box.

Value: $300+

DICK TRACY POLICE CAR: 1951. 10" long. Friction.

This green plastic convertible has a steel base and makes a siren noise. Driving the car is a non-removable plastic Dick Tracy. The toy sold for $1.19 in the 1951 *MONTGOMERY WARD CHRISTMAS CATALOG,* (p.170). The ad shows no markings on the toy and it has been seen that way in person. It has also been seen with Dick Tracy's profile within a badge on the car sides.

The box for the toy has directions for operating it. Hands are illustrated lifting the car and pushing it.

Value: ?

On the left is the 10" long plastic Dick Tracy Police Car which was manufactured in 1951 and operates by friction. On the right is the Dick Tracy Police Station (described in Volume I). Courtesy LLOYD RALSTON GALLERIES, Fairfield, Connecticut.

DICK TRACY RIDER CAR (see after Dick Tracy Squad Car Prototype)

DICK TRACY SQUAD CAR: 1956. 20" long. Friction.

This car was described as "new" in the 1956 *SEARS CHRISTMAS CATALOG* and sold for $3.89.

Riding in the green or blue tin convertible are two white plastic removable figures, driver Dick Tracy and passenger Sam Catchem. The car has a siren and a battery powered flashing signal light in a plastic housing on the left edge near the front of the car.

Value: with box $500-$700

This 20" long tin friction convertible was new in 1956. Inside the car are plastic figures of driver Dick Tracy and passenger Sam Catchem. Courtesy LLOYD RALSTON GALLERIES, Fairfield, Connecticut.

DICK TRACY SQUAD CAR PROTOTYPE:

The Marx Company considered making a hard-top version of the 20" long Dick Tracy Convertible. Well known authority Gene Scala, writing in the *MARX TOY COLLECTOR NEWSLETTER*, describes a prototype in which the bottom is the standard production toy convertible, but the hood has been repainted. The top is from a Fire Car. Each fireman was painted to resemble a policeman and the driver is Dick Tracy. The car is still friction, but a battery operated machine gun has been added in the right side of the windshield. Markings on the prototype are, "Model X4758, Erie, 9/21/56."[2]

Value: ?

DICK TRACY RIDER CAR: 1967. 25" long. Non-powered.

Whether or not this toy was ever produced is uncertain, but it was shown in the 1967 *MARX TOY LINE CATALOG* (p.44) featuring TV promotional items. The toy was described as being a "bottle green blow molded ride-on car with injection molded wheels. Complete with a modern steering wheel with a built in horn and appropriate Dick Tracy markings, this vehicle is sure to be a hit with all little crime stoppers. Complete in every detail – even to the roll bar on the rear fender." The only Dick Tracy marking observed on the car was Dick Tracy's profile within a shield, on the side of the car.

Value: ?

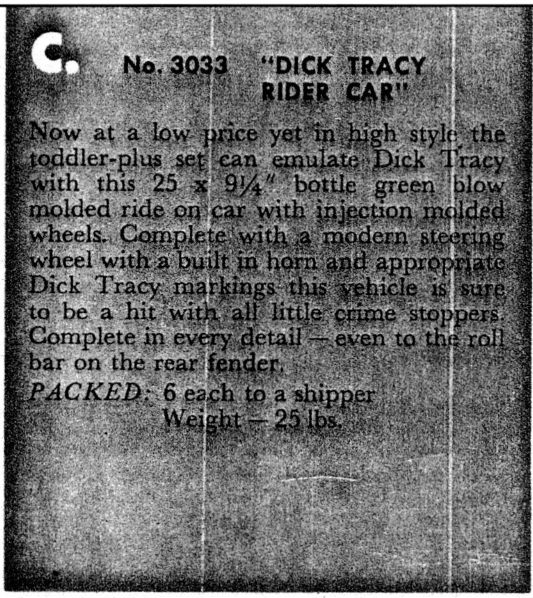
C. No. 3033 "DICK TRACY RIDER CAR"

Now at a low price yet in high style the toddler-plus set can emulate Dick Tracy with this 25 x 9¼" bottle green blow molded ride on car with injection molded wheels. Complete with a modern steering wheel with a built in horn and appropriate Dick Tracy markings this vehicle is sure to be a hit with all little crime stoppers. Complete in every detail — even to the roll bar on the rear fender.

PACKED: 6 each to a shipper
Weight — 25 lbs.

The non-powered Dick Tracy Rider Car measures 20" long. Whether this toy was ever produced is uncertain, but one hopes it was. From the 1967 Marx Toy Line Catalog (p. 44).

DICK TRACY BAGATELLE GAME (see Games section in later volume)

DICK TRACY'S BONNY BRAIDS: 1951. 11" long. Wind-up.

Bonny Braids was the daughter of Dick Tracy and his wife, Tess Trueheart. Joel Smilgis, long-time Editor of *BOX TOP BONANZA MAGAZINE*, provided the above information and, in addition, stated that Bonny Braids was introduced in the comic strip in 1951.

Artist Roy Ahlgren, a former Marx employee, states that Marx manufactured the toy without permission and settled after making a big profit.[3]

In 1950, Marx introduced a "Crawling Baby" which was to become the model for Bonny Braids. One variation of the "Crawling Baby" has been seen with blonde hair, dressed in a spotted red and white blouse and red pants, while another with the same hair color wears a spotted blue and white blouse and blue pants. The hair area of the baby's head is colored yellow and around the edges of the head are embossed hair strands. The baby has painted blue eyes.

The crawling action is truly quite realistic with the arms and legs of the baby pivoting in different directions at the same time. The key for the wind-up mechanism is under the doll's stomach.

Bonny Braids is almost identical in appearance to the "Crawling Baby" described above. She wears the same red and white spotted jacket and pants and she was probably sold with the blue and white outfit, too. The baby does not actually have braids, merely two little tufts of hair, one on each side of her head. Since Marx already had the "Crawling Baby" on the market, in typical thrifty Marx fashion, he merely drilled two holes in the baby's head and inserted strands of hair which stuck out on each side. Presto! Bonny Braids - a new doll.

However, it was necessary to design a new box for the doll. Boxes are always desirable for collectors to have, but in the case of Bonny Braids, the box is particularly important. Nowhere on the doll itself is there any clue that she is a Dick Tracy character. It is only on the red and blue box that the identification can be seen: "Dick Tracy's Bonny Braids From The World Famous Dick Tracy Comic Strip." Below these words is a nice realistic drawing of the doll. On the left side of the doll's drawing, illustrated in the same style as the comic strip, are the proud, smiling parents, Dick Tracy and Tess Trueheart. They are not identified by name. Underneath the baby are the copyright symbol and the words "The Chicago Tribune" and the Marx name and address. At the bottom of the box, in large white letters, are the words, "Sensational Realistic Creeping Action."

The sides of the box show much the same wording and illustrations, except that Bonny Braids is now described as "The Lovable Baby Of The Famous Comic Strip." The end flaps of the box show a picture of Junior (another one of the comic strip characters) and repeat the words "Dick Tracy's Bonny Braids." Wording repeats yet again "She Creeps."

Marx baby dolls were apparently not too successful, for the Marx Company made only a few of them. While the Bonny Braids doll is attractive and does live up to the wording on the box "Realistic Creeping Action," still it is easy to conjecture that little girls would be less than thrilled by the large key protruding from the doll's stomach. Even less to their liking might have been the steel shafts and internal workings of the doll that could easily be glimpsed under her clothing.

Value: Excellent with box $300-$350

A 1951 plastic wind-up Bonny Braids and box (left) along with a 1950 Crawling Baby. Both dolls measure 12" long. Charles & Cheryl Weber collection, R. Grubb photo.

DICK TRACY TALKING TELEPHONE: 1967. 7 3/4" long x 4 3/4" high. Battery operated.

When the receiver carriage buttom is pushed, the plastic telephone plays one of nine randomly chosen phrases, such as, "Junior, change your disguise."

The black telephone base bears a large rectangular decal with Dick Tracy's name in red letters on a yellow background edged with black and white. The white speaker has a decal of Dick Tracy enclosed in a circle.

The colorful box for the toy has a head and shoulders view of Tracy and the words "Dick Tracy Talking Telephone by Marx" Below, in larger letters, is "Dick Tracy's Telephone."

Value: $350-$500

NORA THE NURSEMAID: 1951. Inclined ramp 12" long, Nora walking figure 3" tall, Baby (Bonnie Braids) 1 3/8" tall.

In this unique toy, a plastic Nora the Nursemaid pushes a baby carriage down a metal incline with a rubber baby, Bonnie Braids, in the carriage. The top of the ramp is lithographed with the profiles of Dick Tracy, his wife, Tess Trueheart, and a full face picture of baby Bonny Braids.

The ramp is designed to resemble a pavement. In the middle of it is a comic strip bubble with the words "Watch Bonny Braids' Nursemaid Take Her For A Stroll." At the bottom right of the ramp, B.O. Plenty, another character from the Dick Tracy comic strip, is shown saying, "Thar's A Doggone Purty Baby" (sic). The box for the toy reads "Nora The Nursemaid" and has a drawing of her. Above are the words, "Watch Her Stroll Thru (sic) The Park."[4]

The identical toy was sold under the name of the Charmore Company, this company also being owned by Marx. Although the toys appear to be the same, the boxes are different. The Charmore box does not mention "Nora" at all and has the lengthy title, "Watch The Nursemaid Take Dick Tracy's

Bonny Braids For A Stroll." The words "Dick Tracy's Bonny Braids" are enclosed in a comic strip bubble. On the left side of the box, Dick Tracy, Tess Trueheart and Junior are pictured. On the right side, drawn in larger scale, is Bonny Braids. Under the title, drawn as if in the distance, is a tiny drawing of the nursemaid wheeling Bonny Braids in her carriage.

Value: ?

DORSEY, TOMMY (see Trombone Player).

ESSO TIGER:

The Esso Tiger was introduced in 1962 by the Humble Oil & Refining Company and the slogan "Put A Tiger In Your Tank" became popular. In 1972, the Exxon name was introduced and the tiger re-appeared.

There is some confusion among toy collectors between the "Esso Tiger" and "Tony The Tiger," the character that advertises certain Kellogg's Cereals. While Marx avoided the use of the name "Esso" in naming the various tiger toys, no doubt to evade copyright/royalty problems, it is clear that the Marx Company did manufacture Esso Tiger toys. At least one tiger toy was made holding a straw hat similar to that of the Esso Tiger in the ads. In another version, the box drawing for the tiger clearly shows a striking resemblance to the advertising drawing of the Esso Tiger.

It is not known if Marx ever actually manufactured a Tony The Tiger toy. Tony wears a red neckerchief around his neck with his name "Tony" on it. Nothing like the above description has been seen, but even if Marx left this feature out to avoid copyright problems, the art style of the Esso Tiger, as reflected in the toys' boxes, is softer and more flowing than the more angular style of Tony The Tiger.

So why do people continue to identify the Esso Tiger as Tony The Tiger? The answer may lie in the fact that "Tony The Tiger" is an easier name to remember and more pleasant to the ear than the "Esso Tiger," not to mention the "Exxon Tiger."

ESSO TIGER: 1960s. 11" tall. Battery operated.

This remote controlled orange and white tiger has a white chest. His body and even his remote controller are covered in a tiger-stripe plush, while his head is vinyl.

With his left hand, the tiger tips his hat and in his right hand, he holds a bouquet of flowers. He growls, moves his arms, and walks as well.

Value: $250-$300

ESSO TIGER - FIRST MINIATURE VERSION: 1960s. 8" high. Wind-up.

The miniature version of the tiger has a vinyl head and tin feet. It is similar to the larger version in appearance and action. This tiger also carries a bouquet of flowers in his right hand, but unlike the larger version, he does not hold a hat. That the tiger is not merely missing a hat is shown by the box illustration which depicts him without one, just holding the flowers. As mentioned above, the drawing on the box closely resembles those of the oil company's Esso Tiger ads. The box

An 8" windup Esso Tiger was manufactured in the 1960s. He is similar to the larger battery operated version of the tiger. NEW ENGLAND AUCTION GALLERY, W. Peabody, Massachusetts.

Value: with box $175-$200

title for the toy is "Walking Tiger." The miniature version is seen more frequently than the larger version.

Value: Excellent $75-$100

ESSO TIGER – SECOND MINIATURE VERSION: 1960s. Approximately
7 1/2" high. Wind-up.

Unlike the first miniature version described above, this second miniature version is made of plastic. The orange, black and white tiger does not hold anything since he stands with his paws on his hips in a rather bow-legged stance. The toy moves in a vibrating fashion. This version is seen less frequently than the plush version.

Value: $25-$50

"ESSO" TIGER IN CAR: Mid-1960s. 6 1/2" long. Wind-up.

This is probably the most handsome of all the tiger toys, both because of its detailed lithography and its stunning colors. The toy is completely tin, except for the tiger's vinyl ears.

The tiger is, of course, orange, black and white and sits in his car with his left arm stretched out in front of him. The interior of the car and the steering wheel are red and black. The car's exterior is bright blue and is decorated all over with tigers. The radiator shows the tiger's head and paws, while the hood has part of his body. The sides have full length depictions of tigers and, in a nice touch, each tiger's tail is wound around a car light. The tiger driver's tail dips into the gas tank to show that there is, indeed, a tiger in this tank. (The Esso slogan was "Keep A Tiger In Your Tank.") The license plate reads "Tiger 66," possibly the date the toy was made.

The car the tiger drives is an identical stamping to that of the 1950s Mickey The Driver and Donald The Driver cars (see Volume I) as well as the Careful Johnnie and Dottie The Driver cars (see Volume II). In addition to similar shapes, all these cars – the Tiger car included – have the motor-carriage lithographed onto the bottom of the cars.

One further aspect all the cars have in common is that the heads of the drivers are mounted on springs and shake when the cars move.

The major difference between the Tiger car and the aforementioned cars is that the other drivers are all plastic, whereas the tiger is metal. Another difference is that the other cars were made in the United States whereas the Tiger car, as marked on the base in addition to the Marx logo, was "Made In Japan."

Value: $150-$200

The "Esso" Tiger Car, measuring 6 1/2" long, was manufactured in the mid-1960s. The car is all tin except for the tiger's vinyl ears. Samuel Caley collection, R. Grubb photo.

"ESSO" TIGER ON TRICYCLE: 1960s. 4" tall. Wind-up.

Plastic tiger peddles a tin tricycle in a circle as a bell rings and the top spins. "Tiger" is on the front of the bicycle, directly under the handlebars. Fred and Wilma Flintstone, plus Dino the Dinosaur, also peddle the same kind of tricycle.

Value of Tiger Tricycle: Excellent with box $125-$175.

FUNNY TIGER: 1960s. 7" tall. Wind-up.

It is uncertain whether the Funny Tiger was intended by Marx to resemble the Esso Tiger or even Tony The Tiger. The toy has no identifiable feature to denote a particular resemblence. The tiger lacks the noticeable white chest of the other tigers, but this feature is depicted on the toy in the box illustration. The box illustration resembles the Esso Tiger somewhat more than the actual toy does.

The all tin tiger is orange, black and white. He wears a material tiger skin over his body and also a material bow-tie around his neck. With rubber drum sticks, he beats on a tin drum marked "Funny Tiger." The drum is also covered with a tiger pattern. "Funny Tiger" is repeated on the toy's box, which shows the tiger happily beating his drum.

Value: Excellent with box $175

TIGER GO CART: 1960s. 6" long.

Aside from the fact that it exists, little is known about this toy. The Tiger has the usual orange, black and white coloring, and the Go Cart is light blue, red and yellow with red and black wheels. The handlebars are rather high. Mickey, Donald and Pluto ride similar go carts.

Value: with box $150-$200

THE FLINTSTONES:

Created by Hanna-Barbera Productions (Bill Hanna and Joe Barbera), the Flintstones was not only the longest running animated cartoon in television prime time, it was also the first one. The series was introduced in September 1960. Characters in the family consist of Fred and Wilma Flintstone, their children Pebbles and Bamm-Bamm, and their pet dinosaur Dino. Their close friends are Barney and Betty Rubble. All live in a community called Bedrock. It is believed that Marx or Linemar manufactured toys of all the characters except the children (Pebbles and Bamm-Bamm – one toy was made with Pebbles – see later in this chapter). A Playset of the Flintstones was also made as well as a set of small plastic figures.

DINO THE DINOSAUR FROM THE FLINTSTONES: 1960s. 9" long. Wind-up.

The all tin Flintstone pet, Dino, manufactured by Linemar, is purple with black spots and a pink snout. He walks upright and makes a clicking sound while his mouth moves.

The box for the toy reads "Dino The Dinosaur From The Flintstones." The box illustration shows Dino flanked on the left by Barney Rubble and on the right by Fred Flintstone. All three characters look very happy, almost as if they know how popular they have become and how much the prices of Flintstone toys have risen. Dino The Dinosaur is one of the less common Flintstone toys and the box is even harder to find.

Value: Excellent $450-$500
Excellent with box $1,400-$1,500

FLINTSTONE's BEDROCK EXPRESS: 1962. 26" long x 22" wide. Wind-up.

A very nice toy, the Flintstone Bedrock Express comes with a three-dimensional plastic layout representing the Bedrock settlement. Such items as houses and rocks protrude from the surface and the set also comes with separate trees and rocks.

The layout includes running track for the plastic Fred and Barney Handcar. The handcar measures 5 1/2" long and is shaped like a log. Fred and Barney move back and forth as if working the handcar.

The box for the toy reads "The Flintstone's Bedrock Express." Underneath is "Handcar/Plastic Scenic Track Layout/Trees, Houses/The Works." The box has an attractive illustration showing Fred and Barney working the handcar while all the characters in the series watch them. In the foreground can be seen a car resembling the Marx Flintstone Flivver (see below).

The box bears a different slogan than the usual "One Of The Many Marx Toys/Do You Have All Of Them?" and, instead, has a slogan stating "Another Great Mechanical Toy...By Marx."

Value: Excellent with box $450-$550

Manufactured by Linemar in the 1960s. The tin Dino measures 9" long. Courtesy Smith House, Eliot, Maine.

The box for Dino the Dinosaur is hard to find. Courtesy Smith House, Eliot, Maine.

FLINTSTONE BEDROCK EXPRESS (TRAIN): 1960s. 12" long. Wind-up.

Lithographed in a blue stone-like pattern, the tin train depicts all the various Flintstone characters looking out of the train windows. Fred rides in the locomotive, on the side of which in white letters is written "Bedrock Express." On one of the three cars, as if carved in stone, all in red letters is "Choo choo" while another car in a similar style has "Train."

The box for the toy is marked "Flintstone Bedrock Express" and "Zigzag Action." The train is depicted as travelling through a region with stone dwellings in the background, probably Bedrock.

Value: with box $500-$1,000

Wind-up manufactured in the 1960s, the tin Flintstone Bedrock Express is 12" long. Courtesy Smith House, Eliot, Maine.

FLINTSTONE CARS: 1962. 4 1/2" long each. Friction.

These cars are tin convertibles, lithographed in a blue, white and black stone-like pattern. The wheels are rubber. The same car was made with either a black-haired, red outfitted Fred Flintstone figure, or a red-haired, white dressed Wilma Flintstone figure. The figures have vinyl heads and tin bodies, and are rather large in comparison to the cars. The Wilma car is found more often than the Fred version. Identical cars were also used with another Hanna-Barbera family, Jane and George Jetson. Fred and Wilma came in similar boxes. Betty Rubble, a family friend of the Flintstones, also drives a similar car as does Barney Rubble.

Value: with box $500-$550 each

Fred and Wilma Flintstone each drive 4 1/2" long friction convertibles. Courtesy Smith House, Eliot, Maine.

Betty Rubble drives a similar car to that of Fred and Wilma. Courtesy NEW ENGLAND AUCTION GALLERY, W. Peabody, Massachusetts.

FLINTSTONE FLIVVER: 1962. 7" long. Battery or Friction.

A good Flintstone toy, the tin Flivver came in both a remote controlled battery version and a friction version. It consists of two log-shaped stone-patterned rollers held together by tree branches. On the side of the branches in the friction version is "Flintstone Flivver." The Flivver has a windshield in front and a window in back. The rear extends to a canopy over Fred Flintstone.

In the battery version, a plastic full figure of Fred Flintstone drives the Flivver, while in the friction version, a tin half-figure of Fred with a vinyl head drives. On the base of the friction Flivver, various Flintstone characters are outlined.

The original box for the battery version reads "Fred Flintstone Flivver" and shows Fred driving the vehicle. A circular inset shows a hand operating the controls. The box for the friction version has not been seen.

Value: Battery version with box $1,500-$1,700
Battery version $600+
Friction version $600-$700

In the friction version of the Flintstone Flivver, a tin half figure of Fred with a vinyl head drives the flivver. Courtesy NEW ENGLAND AUCTION GALLERY, W. Peabody, Massachusetts.

The tin remote controlled battery Flintstone Flivver was manufactured circa 1962 and measures 7" long. A plastic full figure of Fred drives the flivver. Courtesy Smith House, Eliot, Maine.

FLINTSTONE LOG CAR: 1977. 5" long. Friction.

This late plastic toy lives up to its name. The body of the car is shaped as if made out of a log, with a seat area cut out for Fred Flintstone. Fred leans back against a seat which extends into a canopy over his head. The plastic wheels, which look as if they are chipped off the edge of a log, are attached to tree branches which in turn are attached to the car.

The box for the toy reads "Flintstone Log Car" and, in smaller letters, "Featuring Fred Flintstone." Fred is shown driving the car in a rather barren landscape.

Value: Mint in box $45-$100

FLINTSTONE PALS ON DINO: 1962. 8 1/2" long. Wind-up.

The tin Dino is purple with black spots and a pink snout. The same dinosaur came in a version with either Barney Rubble or Fred Flintstone riding him, while he walks forward. Both Barney and Fred are tin with vinyl heads. Barney has blonde hair and is dressed in a brown outfit, while Fred has dark hair and wears an orange outfit with dark spots and a blue tie around his neck.

The Barney and Fred versions came in similar boxes with the title "Flintstone Pals on Dino." A smiling Barney is illustrated riding Dino, while the dinosaur's head is turned, smiling back at him. Fred is shown riding Dino, who runs so fast that Barney, who is running on the ground, struggles to keep up.

The version with Fred Flintstone came in a less common box marked Linemar. (The other box is marked Marx.) The rare box is featured in an article by Carol Turpen, "The Creations of Hanna-Barbera" in the November 1991 issue of *ANTIQUE TOY WORLD*.[5] The box shows Fred riding Dino, who is heading towards a house, probably Fred's home. Whether a similar box was used for Barney, too, is not mentioned in the article. The toys are fairly easy to find, but the boxes are not.

Also see larger version later in this section (Fred Flintstone on Dino).

Value: Excellent with box $700-$800 each

These 1962 Flintstone Pals On Dino are wind-ups measuring 8 1/2" long each. The toys are metal except for their vinyl heads. Courtesy Smith House, Eliot, Maine.

FLINTSTONE TRICYCLE: 1962. 4" high. Wind-up.

The Linemar-made red tin tricycle has blue stone-like patterned wheels and the front is lithographed to look like wood. As the tricycle moves, the bell rings and the top of it revolves. The tricycle moves in circles.

The same tricycle was made in versions with three different celluloid figures, with painted features – namely, Dino the Dinosaur, Fred Flintstone and Wilma Flintstone. The same very attractive box was used for all versions, but on the box side, the name for the specific character in the box was indicated. Fred and Wilma are shown riding their tricycles on the sandy ground against a nice blue sky. In the background is a palm tree, the green of its leaves matching Wilma's dress.

Value: Excellent with box $400-$800

A celluloid Wilma rides a tin tricycle. The toy was made by Linemar in 1962 and measures 4" high. Courtesy NEW ENGLAND AUCTION GALLERY, W. Peabody, Massachusetts.

FLINTSTONE TURNOVER TANK: 1960s. 4" long. Wind-up.

This tin tank manufactured by Linemar is lithographed with a stone-like pattern plus various characters from the Flintstone series. A flat figure of Fred Flintstone lifts up the tank from underneath and turns it over. Similar Turnover Tanks were made with the characters of Casper the Ghost, the Jetsons and Popeye.

Value: with box $500-$600

FLINTSTONE TURNOVER TANK – MEXICAN VERSION: 1960s? 4" long. Wind-up.

In addition to the Linemar version, another version of the Flintstone Turnover Tank was also made by Plastimarx, Marx's Mexican subsidiary. This very rare version was shown in the *NEW ENGLAND AUCTION GALLERY CATALOG* for September 1991.

Unlike the standard Linemar Flintstone Turnover Tank, the Mexican version features an angry, sneering Fred Flintstone on the underside turning over the wooden-wheeled tank. Fred is also depicted on the side of the tank, and he does not look too happy there either as he pushes something, a nail perhaps, into the tank. A rather serious Barney pulls a piece of string near a window in which sits a plant. The rest of the characters are happier. Wilma on the hull is actually smiling. Betty is shown on the back and a dog is seen on the turret. All over the tank is a stone-like pattern.

Value: Excellent $1,000-$1,500

FLINTSTONE TRUCK: 1960s.

Also seen in the article mentioned previously (see Flintstone Pals on Dino) was a very rare Linemar Flintstone Truck.[6] Unfortunately, no description is given of this exciting toy, but it appears to be made of tin and is of the tractor-trailer type. Fred Flintstone is lithographed playing a primitive piano, while Barney and Betty Rubble and Fred's wife, Wilma, all watch him play, apparently liking what they hear. Behind Fred is Dino the Dinosaur, also enjoying the music. Behind the others, a sabre-toothed tiger gazing mildly at the proceedings, showing that perhaps "Music doth sooth the savage beast."

Value: $700-$800
with box $3,000-3,500

FRED AND WILMA FLINTSTONE ON DINO: 1960s. 4" long. Non-powered.

This is a plastic toy in which Fred and Wilma Flintstone ride Dino the Dinosaur, who walks down an inclined ramp. The toy will also walk on a flat surface like a table if a weighted string is attached to the toy's front and the string is dangled over the table edge.

Fred is dressed in a yellow garment with black spots and Wilma wears a white dress. Basically, they appear similar to the way they usually look. The real surprise here is Dino, who is not his usual purple color, but instead is predominantly green.

Value: $50-$75

PEBBLES FLINTSTONE ON DINO: 1960s. 4" long. Non-powered.

A small plastic Pebbles rides a rather large Dino down an inclined ramp. Pebbles has red hair and wears a yellow outfit, while Dino is his usual purple with dark spots. This ramp walker is one of the less common versions.
For a photo, see "Ramp Walkers," upper right, page 165.
Value: $50-$75

FRED FLINTSTONE AND BARNEY RUBBLE WALKER: 1960s. Approximately 3 1/2" high. Non-powered.

The toy is the same in action as the previous toy, Fred and Wilma On Dino. This plastic toy features, instead, Barney Rubble and Fred Flintstone. Fred walks behind Barney, with his right arm attached to Barney's shoulder. Both characters walk closely together, heads tilted slightly upward. Fred wears his usual yellow outfit with black spots and Barney wears brown clothes. (For a drawing of this toy see the section on Roy Ahlgren in the chapter "The Marx Company Remembered.")

Value: $50-$75

See also Ramp Walkers in Disney section.

FRED FLINTSTONE ON DINO: 1962. 18" long. Battery powered.

This impressive toy is the large version of the tin windup described earlier (see Flintstone Pals on Dino). It is the big-

The 1962 battery operated Fred Flintstone on Dino measures 18" long. Walking Dino moves his head, neck, mouth and tail, and Fred also moves. Courtesy LLOYD RALSTON GALLERIES, Fairfield, Connecticut.

gest of all the Flintstone toys. Both Sears and Montgomery Ward advertised this toy in their 1962 Christmas Catalogs, but Sears sold it at the significantly cheaper price of $7.44 while Montgomery Ward sold it for $9.77. The Montgomery Ward Catalog described the toy as follows: "Dino the Friendly Dinosaur with TV's [sic] famous Fred Flintstone as 'jockey.' Moves along floor and changes direction when stopped by obstacle. Dino's head rocks, he makes 'dinosaur sound' [whistling] as Fred guides him along...sturdy metal frame covered in soft purple plush. Dino and Fred have vinyl heads." Actually, more than only Dino's head rocks. In his fine book *BATTERY TOYS*, Brian Moran states that the dinosaur also has moving feet, tail, neck and mouth.[7]

Fred sits in a wood-patterned howdah attached to Dino by a belt. When walking Dino moves, Fred also moves.

As stated previously, Dino is purple with black spots, has a purple head and a pink snout. Unfortunately, his coat is often found in faded condition. Fred wears his usual yellow outfit with black spots.

Luckily, it is not too difficult to find the toy, but it is very difficult to find the box. The yellow box is billed "Fred Flintstone on Dino by Marx." A smiling Fred is illustrated riding Dino as the dinosaur walks forward, a contented expression on his face.

Value: $350-$400
With box $1,000-$1,400

MARX'S DINOSAUR, SPACE SHIP INCLUDED IN BIG TV CAMPAIGN

Among the five toys that are being heavily promoted in an extensive Louis Marx television campaign are Dino with Fred Flintstone (#J-9073) and Mystery Space Ship (#3259/60).

A colorful battery-operated item, Dino is a happy-looking dinosaur who carries a

miniature Fred Flintstone on his back. Made of sturdy metal construction, Dino comes in purple plush, complete with spots. Both Dino and Fred have vinyl heads that move forward and backward realistically when the toy is operated.

Fred sits atop Dino in a four-poster cage and "guides" the dinosaur along with a single rein. As the item moves across the floor, both Dino and Fred rock trickily and a "Dino" sound may be heard. The item comes in a box at three pieces per shipping carton.

Description of Fred Flintstone on Dino printed in the October 1962 edition of Playthings, p. 97.

HOPPING DINO, BARNEY RUBBLE, FRED FLINTSTONE: 1960s. 3 1/2" high. Wind-up.

This hopping toy comes in three different versions, featuring either Dino the Dinosaur, Barney Rubble or Fred Flintstone. To tell the truth, the toys are rather frog-like in appearance, what with their long feet looking like flippers, the shape being necessary to maintain their balance when they hop. The frog-like feeling is continued in the toys' bent over postures, probably also for balance. The arms of the characters are lithographed. Some of the characters from the Jetsons were also made is this series.

Value: $500-$600

RUBBLE'S WRECK: 1960s. About 7" long. Friction.

In this unusual toy, a tin Barney Rubble with a vinyl head drives a vehicle lithographed like a log, the side of which is marked "Rubble's Wreck." Two supports in front and two behind Barney Rubble hold a canopy, part of which extends down towards the back of the vehicle.

The tin wheels have the same stone-like pattern as the rollers in the Flintstone Flivver car.

Value: $800-$1000

FRANKENSTEIN'S MONSTER: 1963. 13" high. Battery-operated.

The fantastic movie "Frankenstein," in which the character originated, was produced in 1931. As most people know, Boris Karloff starred in the part of the monster. What many people do not know is that it was the doctor who was named Frankenstein, not his creation.

The 13" battery operated, remote controlled Frankenstein's Monster was manufactured in 1963. Tin and plastic toy bends, moves arms, picks things up and walks. Courtesy LLOYD RALSTON GALLERIES, Fairfield, Connecticut.

Even Sears mistakenly called the monster "Frankenstein" in the 1963 *SEARS CATALOG*, advertising the toy for $5.99. Advertised on the same page are many spectacular Marx toys, including the robot "Mr. Mercury" which has a similar body to that of the monster. Sears described Frankenstein's monster in the following way: "Frankenstein walks again. Bends, moves arms up or down, picks things up, as hands come together by remote control."

This outstanding toy is tin with very well molded green plastic head and hands. The head is an excellent likeness and is properly scary with nails protruding from the neck and blood oozing from the hairline and the mouth. The figure is dressed predominantly in black with a grey shirt. Even the folds in the shirt are shown.

Value: $750-$1,200

FRANKENSTEIN'S MONSTER: 1960s. 6" high. Wind-up.

This toy is the miniature version of the battery operated Frankenstein's Monster. The toy is similar in appearance to the larger version, but is not as detailed. For instance, this monster lacks the folds in his shirt. Another big difference is found in the legs which are much thinner in the smaller version, almost match-stick like. The feet also differ, with the battery operated version having rather thick brown feet while the miniature version has flatter shaped green feet with finger-like pieces protruding from the inner sides. Brian Moran describes this toy as coming in two versions, one mostly tin, the other plastic.[8]

Value: $325-$350

GODZILLA: 1950s. 9" long (including tail), 5" high. Wind-up.

Godzilla, a fire-breathing giant lizard, first made his appearance in the 1956 Japanese film, "Godzilla, King of the Monsters." There were various other Godzilla films, all the way up to 1985, none as good as the original.

It is felt by many that this toy is supposed to be Godzilla, although the box for the toy merely reads "Prehistoric Animal." However, the toy does resemble the monster. It was manufactured by Linemar.

"Godzilla" comes in two color variations, the first being predominantly green and the other mostly reddish-brown. On both versions, the chest and stomach areas are lighter than the rest of their bodies. Both are covered with a pattern intended to similate lizard skin. In action, as the lizard walks forward in a standing position, he bellows while opening and closing his mouth.

The box for the toy shows a realistic drawing of the reddish-brown lizard standing and looking backwards. In the background are a few sparse black lines indicating the landscape. Also on the box is the number "J-4423." The side of the box shows "Godzilla" prone among rocky terrain.

Value: Excellent $250-$350
with box $450-$500

Tin "Godzilla," a 1950s wind-up by Linemar, measures 9' long (including tail) and 5" high. The monster comes in either green or reddish-brown versions. E. Owens collection.

The miniature version of Frankenstein's Monster is 6" high and was manufactured in the 1960s. Courtesy Smith House, Eliot, Maine.

HENRY: 1950s. 5 3/4" tall.

The 1930s comic strip "Henry" was created by Carl Anderson. Henry is a bald-headed little boy who never speaks and who dresses in a T-shirt and short pants.

Made by Linemar, the toy Henry wears a short-sleeved red T-shirt and short blue pants. The Linemar logo is on the back edge of his pants. On the left side of his pants is a paper label with the copyright symbol and "King Features Syndicate." As in the comic strip, the little boy is bald and his features are shown by a few simple lines, a style frequently used by Linemar for other toys.

In action, Henry's movable head goes up and down as he revolves in a circle, while he licks his candy cane with a felt tongue. The tongue is so frequently missing that it is not commonly realized that he is supposed to have one. However, the lack of the felt tongue results in the toy having wear on its face as it bangs into the candy cane.

Value: with box $500-$600

Henry was manufactured in the 1950s by Linemar and is 5 3/4" high. His head moves up and down as he licks a candy cane and revolves in a circle. Samuel Caley collection, R. Grubb photo.

HERMAN MOUSE: 1950s. 5" high. Wind-up.

Herman Mouse and Katnip the Cat may have been created to cash in on the success of the Hanna-Barbera cartoons of Tom and Jerry. Obviously, if this was the case, these hopes were not realized. The cartoons were shown in the 1950s and 1960s and were also featured in Harvey Famous Comics.

Manufactured by Linemar, the tin toy Herman Mouse wears a long-sleeved shirt and pants and his arms are lithographed close to his body. The mouse has rather long ears and the facial features are nicely defined. He also wears a cap and bow-tie. When activated, Herman vibrates and his head, which is mounted on a spring, shakes.

Although appealing, Herman is one of the few Linemar character toys that can be bought reasonably. The toy is rare, but most people do not recognize the little mouse.

Herman comes in a box which shows the toy with his arms outstretched. The box's title is simply "Herman" and the bottom has the Linemar logo and copyrights for both Linemar and Harvey Famous Comics. Also on the box is "Linemar/Best By Far." Another box illustration has been seen which more closely resembles the actual toy, in that Herman's arms are shown close to his sides. Unlike in the other illustration, Herman's tail cannot be seen.

Value: $250-$350

SEE ALSO KATNIP CAT.

HOWDY DOODY:

The famous children's star "Howdy Doody" was created by "Buffalo" Bob Smith and ran on television for a long time, from December 1947 to September 1960. Besides the star, Hoody Doody, there were various other characters including Clarabell the Clown. Howdy Doody and Clarabell the Clown are believed to be the only tin figures (all versions of which were made by Linemar), although some of the other characters were made in plastic, non-moving figures.

CLARABELL DELIVERY WAGON: 1950s. Friction.

Not much is known about this toy. The toy is all tin except for Clarabell's celluloid head. The clown peddles a unicycle, to which is attached a small two-wheeled wagon.

Value: $350-$500

CLARABELL STANDING ON HANDS: 1950s. 5" high. Wind-up.

This striking toy moves with a vibrating motion while walking on his hands. The manufacturer Chein also made a similar toy.

As befits a clown, Clarabell wears clown makeup and has red hair. His clown suit is designed in a red and white striped pattern.

The box for the toy reads "Clarabell Standing On Hands" and shows the clown doing just that. The stripes on the illustration's figure are not nearly as prominent as on the actual toy.

Value: with box $500-$600

HOWDY DOODY PAL CLARABELL: 1950s. 7" high. Remote control cable action.

This tin figure of Clarabell walks in an upright position when the handle of the cable attached to the clown is squeezed. The problem with cable action toys is that the hand tends to become somewhat tired squeezing the handle.

The appearance of this Clarabell resembles that of the other toys, but the head seems a bit bigger and the toy is lithographed in greater detail. In addition, Clarabell has a nameplate attached to his chest.

The box for the toy has a fine illustration of Clarabell's head and shoulders. A very small Howdy Doody is depicted in the background.

Value: $400-$500
Mint with box $850-$1,000

HOWDY DOODY PLAYING BANJO: c. 1950. 5" high. Wind-up.

The title almost says it all. As well as playing the banjo, this tin Howdy Doody moves his head. In his book, *COLLECTING TOYS*, Richard O'Brien is the only person the author has found who mentions this toy, but the above is all the information he has on it.[9]

Value: $500 and up

Manufactured by Linemar, the tin Howdy Doody Pal Clarabell is 7" high. He walks when an attached cable handle is squeezed Courtesy Smith House, Eliot, Maine.

HOWDY DOODY PIANO: 1950s. 4" x 5" base. Wind-up.

This is a splendid toy in which Howdy Doody dances while Clarabell Clown plays the piano. Howdy Doody is mounted on a pole at the left side of the piano.

The toy is a study in contrasting patterns. There is Howdy Doody's plaid shirt and Clarabell's wavy, striped clown suit. The upper half of the piano is spotted and the toy's base has wide stripes.

This toy should not be confused with the Howdy Doody Piano made by Unique Art. That toy is much larger and has "Buffalo" Bob Smith playing the piano instead of Clarabell.

Value: $1200-$1500

This strikingly colored rare tin Wind-up toy was manufactured by Linemar in the 1950s. The base of the toy is 4" x 5". Howdy Doody dances while Clarabell Clown plays the piano. Courtesy NEW ENGLAND AUCTION GALLERY, W. Peabody, Massachusetts.

HUCKLEBERRY HOUND:

The Huckleberry Hound television show, one of the earliest cartoon series by Bill Hanna and Joe Barbera, ran from 1958 to 1962. The character Huckleberry Hound was a dignified, good-natured dog from down South. In 1960, the show won an Emmy award, being the first television cartoon series to achieve such a distinction. In addition to the toys described below, Huckleberry Hound came in various miniature figure sets.

HOPPING HUCKLEBERRY HOUND: 1962. 4" high. Wind-up.

This toy was manufactured by Linemar and has a bent over, frog-like appearance similar to the Hanna-Barbera figures of Quick Draw McGraw, Fred Flintstone, Barney Rubble and Dino the Dinosaur (see the Flintstone toys). The hound wears a lithographed bow-tie and cap.

Value: $150-$175

HUCKLEBERRY AEROPLANE (sic): Late 1950s-early 1960s. 9" long x 9" wingspan. Friction.

This attractive tin airplane with a plastic propeller was manufactured by Linemar and has lithographed Hanna-Barbera characters all over it, including the wings and rudder. It is not certain who some of the figures are, but Yogi Bear is shown on the side of the plane along with the words "Huckleberry Hound" and "Yogi Bear." Behind Yogi, near the back wing, is the Linemar logo. A vinyl-headed Huckleberry Hound with painted features sits in the open cockpit.

The box for the toy shows Huckleberry flying the plane in the sky, with its propeller spinning. As in the actual toy, the box illustration shows various characters lithographed all over the plane.

The same airplane was also used in the Quick Draw McGraw toy "Animal Airplane," but that toy and box were lithographed differently and, of course, the Quick Draw McGraw figure was different.

Value: with box $975-$1,000

Tin Huckleberry Aeroplane manufactured by Linemar in the late 1950s to early 1960s measures 9" long by 9" wingspan. At left is Flintstone Pal Barney Rubble on Dino. Courtesy LLOYD RALSTON GALLERIES, Fairfield, Connecticut.

HUCKLEBERRY HOUND CAR: 1962. 4 1/2" long. Friction.

Huckleberry Hound drives a blue tin convertible with rubber wheels. Huckleberry's blue and light brown movable head is vinyl and he wears a yellow hat. His body is red tin.

Except for the difference in heads and lithography on the cars, this toy is exactly the same as the cars of the Hanna-Barbera characters Fred and Wilma Flintstone or Quick Draw McGraw and Yogi Bear.

The box for the toy reads "Huckleberry Car" and has a rather limited illustration of Huckleberry Hound driving his car plus a few horizontal lines to suggest a road. Markings include the Marx logo and "Made In Japan."

Value: Excellent with box $250-$275

The 1962 friction Huckleberry Hound Car is tin and measures 4 1/2" long. Huckleberry has a vinyl head. Courtesy NEW ENGLAND AUCTION GALLERY, W. Peabody, Massachusetts.

Rare Huckleberry Hound Linemar tin car was made in 1961. The Wind-up measures 9" long. Courtesy NEW ENGLAND AUCTION GALLERY, W. Peabody, Massachusetts.

HUCKLEBERRY HOUND CONVERTIBLE: 1961. 9" long. Wind-up.

Made by Linemar, the rare tin car is dark blue. A blue vinyl-headed Huckleberry with a yellow hat drives the car. The car's interior is lithographed with depictions of Huckleberry, Yogi, and Quick Draw McGraw.

Value: $725-$1,000

HUCKLEBERRY HOUND GO-CART: 1960s. 6" long. Friction.

This Linemar toy is completely tin except for Hucklebery Hound's vinyl head and the go-cart's black rubber wheels. The Linemar logo, "Japan" and "Linemar Co. Inc" are embossed on the toy's base.

The Huckleberry Hound Go-Cart came in an open display box which consists of a base for the cart and a mounting on the side. Just what is written on the box is unknown; the following few words appear on the base: "...with new strong friction motor."

Linemar also made Quick Draw McGraw and Yogi Bear Go-Carts, both probably packaged in the same manner.

Value: with box $250-$300

HUCKLEBERRY HOUND TRICYCLE: 1961. 4" high. Wind-up.

This toy is similar to the Flintstone Tricycle described earlier, but the lithography is different. The tin wheels of this red and yellow tin tricycle have a design resembling that of a kaleidoscope. The revolving bell is red and yellow.

Like the Flintstone figures who peddle a similar tricycle, Huckleberry Hound is made of celluloid. He is light blue in color with a tan snout. He wears a yellow hat.

Value: Mint $550-$750

HUCKLEBERRY HOUND MICRO-VIEWER: 1960s. Approximately 3 1/2" long. Button manually operated.

This is an interesting and inexpensive plastic toy which is shaped like a circus wagon. The idea is to look through the window-shaped viewer on the front of the toy while pressing a button on its base. When these actions are performed, a total of eight different scenes of Huckleberry Hound can be viewed.

Viewers with different shapes were also made for the characters Quick Draw McGraw and Top Cat.

Value: $12-$25

HUCKLEBERRY HOUND AND YOGI BEAR WALKER: 1960s. 3 1/2" high. Non-powered.

In this plastic toy, Huckleberry Hound and Yogi Bear walk down an inclined ramp, or to the edge of a table if a weighted string is attached to the toy and dangled over the table edge. Huckleberry is behind Yogi with his right arm attached to Yogi's shoulder. Since both characters are really one toy, they walk very closely together.

Huckleberry is blue with a tan snout and wears a red tie and brown hat. Yogi is brown, also with a tan snout, and wears a green tie and hat. The toy is the same type as the Barney Rubble and Fred Flintstone Walker described earlier.

Value: $50-$75

THE JETSONS:

The space age family, the Jetsons, created by the prolific and talented Bill Hanna and Joe Barbera of Hanna-Barbera Productions, includes parents Jane and George, children Elroy and Judy, plus a robot maid Rosey and a pet dog Astro. Although the television show lasted only a year, from 1962 to 1963, the show's popularity was continued through reruns over the years.

ASTRO WALKER: 1960s. 2 1/2" high. Non-powered.

This plastic toy, the first of three walkers with Astro, the Jetson family's dog, features the canine by itself.

All four of Astro's legs are movable to enable the dog to walk down an inclined ramp, or to a table edge if a weighted string attached to the toy is dangled over the table edge. A loop is attached to the front of the toy, so it can be pulled if a string is inserted.

Astro is primarily grey with a pink snout and a green collar. His red tongue dangles out of his mouth.

Value: $50-$75

ASTRO AND GEORGE JETSON WALKER: 1960s. 3 1/2" high. Non-powered.

Unlike the toy version immediately preceding, this Astro toy has the dog walking in an upright position with only two movable legs. He is smiling and his tongue does not stick out as in the other version.

George Jetson follows Astro and they are attached by their hands, rather than the rear character having his hand on the front character's shoulder as in some other walker toys such as the Flintstones. George also has movable legs, has brown hair and is dressed in white.

Value: $50-$75

ASTRO AND ROSEY THE ROBOT WALKER: 1960s. 3 1/2" high. Non-powered.

The toy walks down an inclined ramp or across a table is a weight is attached to the toy by a string and the weight is dangled over the table side.

Astro follows Rosey the Robot, joined to her arms by his paws. He is the same figure used in the Astro and George Jetson Walker toy and is colored the same grey with pink snout and green collar. Rosey the Robot is predominantly light blue.

Value: $50-$75

HOPPING ASTRO, ELROY AND GEORGE JETSON, ROSEY THE ROBOT: 1960s. Astro, Elroy and George Jetson approximately 3 1/2" high. Rosey the Robot 2" high. Wind-ups.

All of these tin hoppers are similar with their bent-over frog-like appearance, large flipper-like feet and lithographed arms. Astro has an antenna on his tail. Rosey the Robot is bent over almost double and her body sways up and down. The hoppers from the Flintstone series are similar toys.

Value: Astro, Elroy and George Jetson $500-$600 each
Rosey the Robot $700-$800

THE JETSON EXPRESS: 1960s. 13" long. Wind-up.

Lithographed on this tin toy's locomotive and three cars are the various characters, with their names, from the Jetsons series.

The blue-grey locomotive bears the words "Jetson Express" while the three cars each have one word of the label "Jetson/Choo Choo/Train." George Jetson is lithographed on the locomotive, his wife Jane on the first car, Rosey the Robot on the second car and Elroy on the last car.

Value: $400-$500

The tin 1960s windup Jetson Express is 13" long. Lithographed on the train are teh names of various Jetson characters. Courtesy Smith House, Eliot, Maine.

THE JETSON SPACE BALL GAME: 1960s. Size unknown.

The toy is a bagatelle action game in which one shoots for points. The tin base is lithographed with various characters from the series and reads "The Jetsons" and "Space Ball." The base is covered with a plastic top. The toy's box is also lithographed with characters from the series.

Value: Excellent $250-$300

The Jetson Space Ball Game, made in the 1960s, is tin with a plastic cover. Various Jetson characters are lithographed on the tin base. Courtesy NEW ENGLAND AUCTION GALLERY, W. Peabody, Massachusetts.

JETSON TURNOVER TANK: 4" long. Wind-up.

The tin Jetson Turnover Tank is similar to the Flintstone and Casper Turnover Tanks. It is predominantly greenish-blue and has various Jetson characters from the series lithographed all over it. A flat figure of Rosey the Robot lifts the tank from underneath so that it turns over. The tank was made by Linemar.

An unauthorized version of this tank was made in China and measures 5" long, 1" longer than the Linemar version's 4" length. The unauthorized version is blue-green and is marked with the Jetson name on the turret. Various Jetson characters are lithographed on the tank.

Value of the Linemar Tank: Excellent $375-$400

KATNIP CAT: 1950s. Approximately 4 1/2" high. Wind-up.

Katnip Cat, along with Herman Mouse, appeared in the "Herman and Katnip" Harveytoons movies in the 1950s (possibly earlier) and also appeared in the comics and on television.

The Katnip Cat toy was manufactured by Linemar. This rare tin toy is similar to the Linemar Casper the Ghost. The cat has the same pose as Casper, with its lithographed hands resting on its stomach, but the lithography differs. The dies of the two toys are the same except for their heads. Like Casper, Katnip Cat bounces around as his spring-mounted head shakes.

Katnip has a white face with a large red nose, brown body and yellow hands and stomach. Like Herman Mouse, this uncommon toy is not easily recognized.

Value: $250-$350

(Above) Color photo of Katnip Cat is courtesy of NEW ENGLAND AUCTION GALLERY, W. Peabody, Massachusetts. (Left - Black and white photo) The 1950s Linemar tin Katnip Cat measures approximately 4 1/2" high. He bounces around and his head shakes. (Right) Similar toy Casper the Ghost shown for comparison. Courtesy LLOYD RALSTON GALLERIES, Fairfield, Connecticut.

SEE ALSO **HERMAN MOUSE**.

KING KONG (see Mighty Kong)

LITTLE AUDREY: 1950s. 4 1/2" high. Wind-up.

Little Audrey originally appeared in Harvey Comics in 1948. In 1949, the character starred in a series of cartoons which continued into the 1950s.

One look at this Linemar toy and one is instantly reminded of the Casper the Ghost toy, also made by Linemar. Audrey stands in the same pose as Casper, with her lithographed hands on her stomach and her features are drawn in the same way. In the toy version, as in the cartoon, Audrey has retained the Peter Pan collar of her blouse, but her hair has become shorter.

The character's head is mounted to a spring and shakes as the toy moves with a vibrating motion.

Value: $250-$300

THE MAN FROM U.N.C.L.E.:

This popular series was on television from 1964 to 1968. It concerned the secret agent organization of U.N.C.L.E. (United Network Command for Law and Enforcement) fighting the evil crime syndicate, Thrush. Starring as Napoleon Solo was Robert Vaughan, Illya Kuryakin was played by David McCullum and Alexander Waverly was portrayed by Leo C. Carrol.

In 1966, the series "The Girl From U.N.C.L.E." was introduced, but it died a quick death in 1967. Besides the items listed in this chapter, the Marx Company manufactured a Secret Code Wheel Pinball Game and a Man From U.N.C.L.E. Pistol Cane plus a few 6" figures from the series.

THE MAN FROM U.N.C.L.E. COUNTERSPY OUTFIT: 1960s. Box 15" long.

This set contains no action figures but has everything else the well-equipped miniature spy could possibly need, such as a weatherproof trench coat with secret pockets, a bullet shooting lighter plus bullets, an exploding cigarette case, a walkie-talkie, a nerve spray camera, plus a pistol with a stock extension, scope sight, message missile and launcher barrel.

Just to insure that the spy got a chance to use all the great equipment, the outfit came with numerous disguises, including false beards, mustaches, glasses, an eye-patch and make-up. There was also a pair of hand-cuffs for evil-doers and even a cardboard briefcase in which to carry all the equipment. An instruction sheet was included.

The box for the toy shows a picture of Robert Vaughan and David McCullum, plus an illustration of an agent wearing the trench coat. The spy holds the coat open to show the secret pockets.

One side of the box has another photo of Robert Vaughan and illustrates the special equipment and disguises contained within the set. The other side has a photo of David McCullum and shows the "all-purpose counterspy pistol" and all the pieces of equipment that came with it.

Whoever owned this set in the past, or who will do so in the future, will have a lot of fun with it.

It has been rumored that there is a Marx Boy From U.N.C.L.E. outfit, but no information on this toy has surfaced. Nothing much is known of a Man From U.N.C.L.E. Racing Set either, except that its cars collapse when they crash.

Value: $400-$600

THE MAN FROM U.N.C.L.E. MINIATURE LICENSE PLATES: 1967. 4" long x 2" high.

These attractive red and black metal license plates are embossed with black and white letters. Each plate has the symbol of The Man From U.N.C.L.E. on it, which consists of the globe and a figure standing beside it. Underneath are various names in which the license plates were issued, including: "The Girl From U.N.C.L.E." – "The Man From U.N.C.L.E." – "Illya Kuryakin" – "Napoleon Solo."

"The Girl From U.N.C.L.E." is the only one of these license plates to have a female figure (rather than a male figure) standing beside the globe. Also, the style of the lettering in this license plate is wider and stands in contrast to the slimmer straight lettering in the other license plates.

The license plates were sold in bags with the globe and man motif being repeated on the header cards, along with "U.N.C.L.E.," "J-6123" and the Marx logo. The bags are sometimes missing from the plates.

The Marx Company made these license plates with other characters, also, such as "Batman."

Value: $50-$75

MAGIC MARXIE/THE SPORT CAR DRIVER: 1960s. 1 1/2" long. Non-powered.

The Magic Marxie symbol was introduced by Marx in 1959. It consists of a pixie wearing a dapper cap. The Marxie has a head, arms and legs, but the rest of his body is shaped like a Marx logo. The logo consists of the circle and the words "Marx Toys."

This 1960s plastic Magic Marxie car is 1 1/2" long. Although non-powered, the car can move. The same toy came in versions with cars of different colors. The Magic Marxie was also made into a pin. S. Caley collection, R. Grubb photo.

This plastic toy consists of the Magic Marxie driving a sports convertible. The Marxie retains his dapper cap (blue in this instance) and circular body (white) but has no arms or legs. The red car has a yellow interior and grey wheels. Small though it is, the car's wheels can move.

The red box for the toy reads Magic Marxie/The Sport Car Driver and shows an accurate illustration of the toy, except the Marxie is shown wearing a shirt. Also on the box is a circular logo marked "Elm Toys" and "Hong Kong Made."

Value: $50-$75

MARXIE TOMATO SOUP CAN AUTOMOBILE PROTOTYPE: 1960s-1970s. 4" long x 3 1/4" high. Friction.

This plastic prototype consists of an interesting figure resembling a soup can with arms and legs, driving a plastic production auto.

The hand-painted red, white and yellow figure is marked "Marxie Tomato Soup" and its raised head on the tomato soup can is appropriately tomato shaped. It has blue eyes, a round yellow nose and two protruding teeth, and wears white gloves and yellow boots.

The car is blue with a painted gold radiator and black rubber tires. The underside is embossed with "1910 Ford" and the Marx logo with the less common than usual designation "Made In Taiwan."

Value: ?

Plastic prototype Marxie Tomato Soup Can Automobile has a soup can figure driving a plastic production car. The toy measures 4" long x 3 1/4" high. Philip Rolin collection, Robert Grubb photo.

THE MIGHTY KONG:

One of the best known science fiction films ever made, "King Kong" was the story of a giant ape. It was originally produced in 1933 and an inferior version was re-made in 1976.

THE MIGHTY KONG: 1963. 11" tall. Battery operated.

This remote controlled toy is a good representation of the movie character. Kong was advertised in the 1963 *SEARS CHRISTMAS CATALOG* and was described as "Gorilla's head sways, he growls, beats chest, walks. Fur-look [brown] plush covers 11 inch body." The toy sold for $6.79.

The ape has a vinyl face, vinyl hands and tin feet. He wears a metal collar and arm bands linked by a chain.

The toy comes in a handsomely illustrated box which is primarily red and blue. The angry ape, in chains, in shown grabbing a car, while the Empire State Building is cracking and getting ready to fall. An airplane flies overhead. Bold red letters read "The Mighty Kong" while smaller red letters urge "Make Him Walk! / Make Him Beat His Chest And Growl!"

Value: Excellent with box $600-$700

The battery-operated, remote controlled Mighty Kong is 11" tall and was made in 1963. He growls, beats his chest, his head sways and he walks. Courtesy LLOYD RALSTON GALLERIES, Fairfield, Connecticut.

MIGHTY KONG – MINIATURE VERSION: 1960s. 8" tall. Wind-up.

Unlike the miniature versions of many other large versions, this smaller wind-up Mighty Kong is a good substitute for the larger battery operated toy, if one does not want to pay quite as much. The smaller version resembles the larger one quite closely, whereas many other miniature versions lack the detail of their bigger counterparts.

The most significant difference between the large and small sizes, apart from method of operation, is found in the graphics on the respective boxes. The smaller toy comes in a red and black box which merely says "Gorilla" with no reference to "Kong." Just the head of the gorilla is shown, depicted as roaring. Wording on the box states "Permanent key-clock type motor / A realistic plush toy with vinyl head / It growls! / It prowls!" The Marx logo is also shown. While this is an attractive box, the illustration does not compare with the illustration of the larger Kong and the falling Empire State Building.

Value: with box $400-$500

MILLER, GLENN (see Trombone Player)

POPEYE:

The "Popeye" character was created by E.C. Segar in 1929. Before being called "Popeye," Segar's comic strip was known as the "Thimble Theatre" and Olive Oyl had another character as her boyfriend. When Popeye was introduced, his character became so popular that eventually the strip took on his name and centered on his adventures.

In 1933, Popeye made his film debut in a Betty Boop cartoon and went on to star in his own film cartoons. Popeye's television debut was in the late 1950s.

In the past, Marx had made quite a few Popeye toys. Linemar continued this pattern, manufacturing an extremely large number of outstanding Popeye toys. In Marx toys, Popeye usually wears a navy blue top with a red collar, navy blue and white striped trousers, or solid blue trousers, and a navy cap or, less commonly a white cap. In Linemar toys, Popeye wears a black top with a red collar, blue trousers, brown shoes and a blue or white cap or – less commonly – a red cap.

This 8" tall wind-up Gorilla was made in the 1950s and is a smaller version of the battery-operated Mighty Kong. Courtesy LLOYD RALSTON GALLERIES, Fairfield, Connecticut.

This 12" high battery-powered Bubble Blowing Popeye was manufactured by Linemar in the 1950s-1960s. He alternately raises the can of spinach and then the pipe, which lights up, to his mouth. His mouth then opens and blows out bubbles. Courtesy LLOYD RALSTON GALLERIES, Fairfield, Connecticut.

BUBBLE BLOWING POPEYE: 1950s-1960s. 12" high. Battery-operated.

This colorful toy is all tin except for Popeye's plastic pipe. Popeye, dressed in his sailor suit, stands on a blue and yellow base with white and red wording reading "Bubble Blowing Popeye." In his left hand is a can of spinach and in his right is a pipe. When activated, Popeye alternately raises the spinach can and then the pipe, which lights up, to his mouth. His mouth then opens and blows out bubbles.

The hard-to-find box shows Popeye whirling around in frustration, his cap falling off his head as bubbles fly out of his mouth and float all around him. Regarding the scene with pleasure is the baby Swee' Pea (sic), a character from the strip. The side of the box states that "Bubbling liquid may be purchased at any store." Also on the box side is a Linemar slogan which in this instance is the same as the regular Marx slogan except for the name: "One Of The Many Linemar Toys Do You Have All Of Them?" The usual Linemar slogan is "Linemar Best By Far."

Value: $900-$1,000
With box: $1,900-$2,000

JUGGLING POPEYE AND OLIVE OYL: 1950s-1960s. 9" high. Wind-up.

One of the most fantastic Popeye and Olive Oyl toys ever made, this desirable item is one of the highest priced of all. The toy consists of Olive Oyl seated in a twirling chair which is balanced on Popeye's nose. This effect is achieved by a rod which extends from the top of Olive Oyl's head all the way down on Popeye's nose, possibly to the base. In his right hand, Popeye holds a tin stick and as his arms move, the stick hits the chair legs, causing Olive Oyl and the chair to twirl.

Popeye is dressed in his usual sailor suit and Olive Oyl wears a red blouse and black skirt with a wavy blue line near the edge. Unlike Popeye's painted arms, her arms are lithographed at her sides. Attached to her head is a rubber pony-tail.

The yellow box for the toy reads "Juggling Popeye and Olive Oyl" and depicts Olive Oyl with moving arms and a straight line in her skirt, neither of which the actual toy has. Otherwise, the illustration is an accurate representation.

Also on the box is the Linemar logo and copyright symbols for both "Linemar Co. Inc." and "King Features Synd" (sic). In addition, there is the familiar Linemar slogan, "Linemar Best By Far." A similar toy was made by the Japanese manufacturer TPS and called Clown Juggler with Monkey.

Value: Excellent $1,900-$2,000
Excellent with box $3,000-$3,100

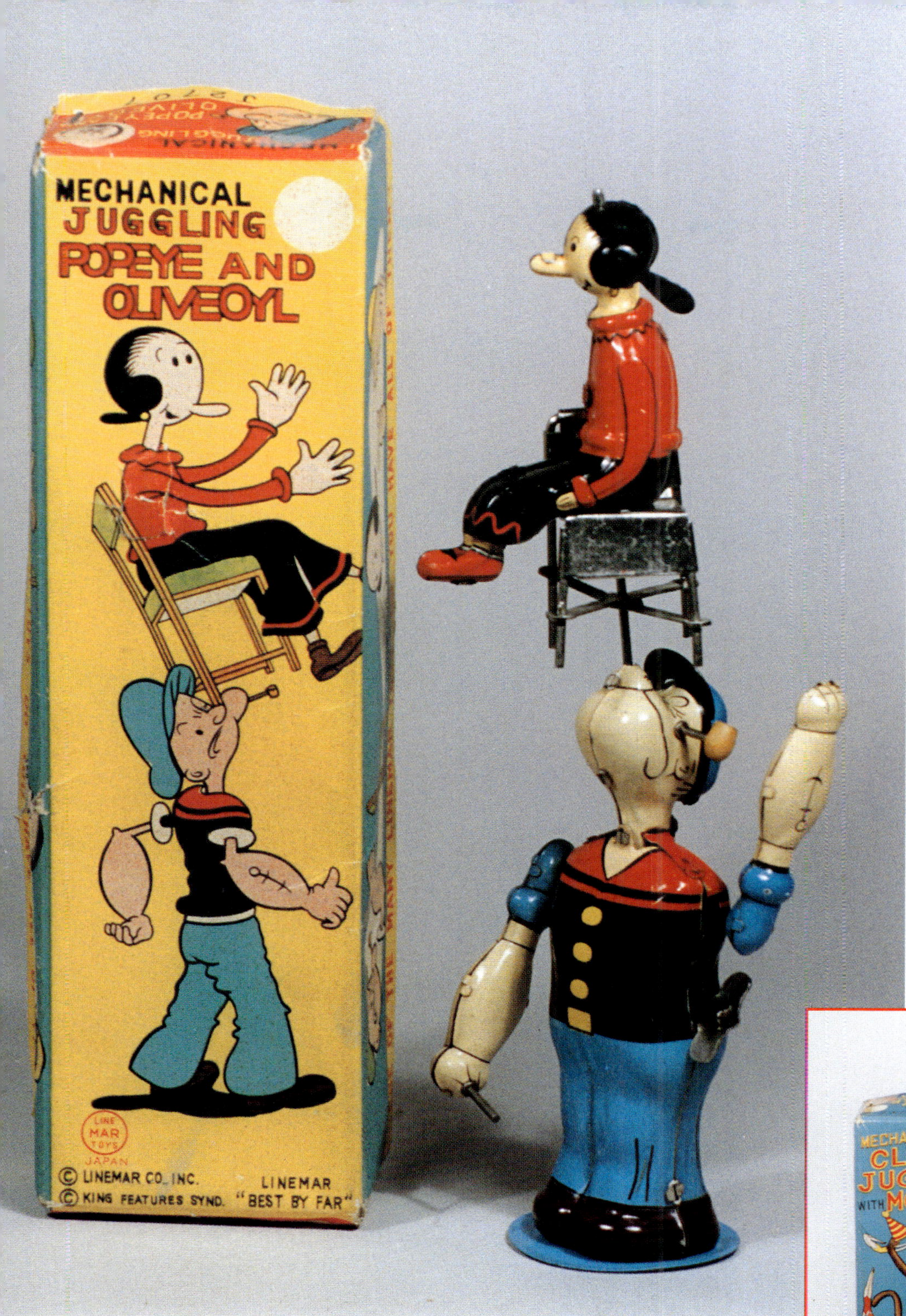

This fantastic 1950s toy, manufactured by Linemar, is a tin windup measuring 9" high. Popeye hits the chair with a stick, which causes Olive Oyl and her chair to twirl around. Courtesy NEW ENGLAND AUCTION GALLERY, W. Peabody, Massachusetts.

Clown Juggler with Monkey, manufactured by the Japanese firm TPS, bears a close resemblance to the Juggling Popeye and Olive Oyl. Courtesy NEW ENGLAND AUCTION GALLERY, W. Peabody, Massachusetts.

OLIVE OYL BALLET DANCER: 1950s-1960s. 5 1/2" high. Friction.

Olive Oyl is tin except for her rubber nose. She spins on her toes when a rod at the edge of her skirt is pulled out.

Olive is brightly dressed in a yellow dotted top with a red rose at each edge and a red striped skirt. To complete the outfit, Olive wears long yellow socks. The Olive Oyl Ballet Dancer has also been seen wearing an olive dress which is very appropriate, considering her name.

The toy comes in such a multi-colored box that it makes even Olive's colorful outfit seem subdued by comparison. The box must include every color, with the exceptions of purple and pink. Olive Oyl is illustrated dancing while higlighted by spotlights and floating notes of music. In the background sits boyfriend Popeye contentedly watching her dance. Seeing Popeye either contented or in the background is not a common sight.

Value: With box $750-$900

POPEYE ACROBAT: 1950s. 13" high. Wind-up.

This exceedingly rare toy consists of a tin Popeye with a jointed body swinging back and forth on a flexible rod which is mounted on a rocking horse. Popeye wears a sailor suit and a blue cap. No sleeves have been lithographed on his rather long arms. The base of the toy reads "Popeye Acrobat" and shows various characters associated with Popeye, such as Olive Oyl, Swee' Pea, Wimpy and Bluto.

The rod and base of the Popeye Acrobat are similar, or even the same, as a pre-1950s toy known as the "Acrobatic Marvel" (see Volume I, page 50). The Acrobatic Marvel is a non-character monkey. Contrasted with the facts that Popeye is a comic character and the toy is very rare, and the price difference in these two otherwise similar toys is several thousand dollars.

The equally rare box for the Popeye Acrobat depicts the sailor flying through the air on a trapeze. Such is the breeze whipped up by his flight that onlooker Swee' Pea's cap is blown off. The flexible rod with Popeye detaches for storage in the box. While this feature is convenient, it accounts for many of these parts being lost.

Value: With box $4,200-$4,500

POPEYE AIR-O-PLANE: [sic] 1950s-1960s. 6 1/4" long. Wind-up.

This very rare all tin Popeye Air-O-Plane is built something like an eccentric car with wings. Like the car, it has two small front wheels which can pivot, and two large back wheels, which cause the plane to move in an erratic, zigzagging manner.

Popeye's head sticks out of the cockpit and he is, as always, smoking his ever-present pipe. This time he does not wear his sailor cap, wearing instead a lithographed helmet and goggles.

The plane is silver, red and yellow with the number "7," a star and "Popeye" marked on the fuselage side. The wings have a star and bar motif and the propeller is silver and red. A simulated rivet design covers the fuselage.

The excessively rare box is titled "Popeye Air-O-Plane" and depicts Popeye flying in a cloud-filled sky.

Value: $1,200-$1,500

POPEYE AND MEAN MAN MECHANICAL FIGHTERS: 1950s-1960s. 4" high. Wind-up.

This is an uncommon toy in which the celluloid figures of Popeye and Bluto are each mounted on separate two-wheeled tin platforms. The platforms are attached to each other by a metal strip. As the toy moves forwards and backwards, Popeye and Bluto bob and shake while their jointed arms swing.

Popeye is dressed in a purplish-blue shirt with a red collar, white cap and blue pants, while Bluto wears a brown shirt with a white collar, white cap and yellow pants. Both figures may have come with different colored clothing. The base is primarily red, yellow and blue.

The Popeye and Mean Man Mechanical Fighters has a base similar to the Marx Ferdinand the Bull and Matador toy, and the Nifty Maggie and Jiggs toy.

The original box for the toy reads "Popeye And Mean Man Mechanical Fighters" and also has the King Features Syndicate copyright, Linemar logo and Linemar slogan "Linemar Best By Far." The amusing illustration on the the very rare box shows Popeye challenging the giant "Mean Man." Although Popeye is smaller, he has powerful bulging muscles, while the "Mean Man's" are tiny and puny.

Value: With box $4,500-$5,000

POPEYE AND OLIVE OYL MECHANICAL TRICYCLES: 1950s. 4" high each. Wind-up.

This toy is similar to the Flintstone Tricycles ridden by Fred or Wilma Flintstone. Popeye or Olive Oyl, with tin bodies and celluloid arms and legs, each peddle a blue, red, yellow and white tin tricycle, which operates in circles. The front of the tricycle has a flower design and the tin tires are striped. As the bicycle moves, a bell rings and the top of it revolves.

Popeye wears his sailor suit and a white cap with a light blue brim. Olive Oyl wears a red blouse and a black skirt.

Although the box for Popeye has not been seen, the box for Olive Oyl has. Since these two toys are so similar, it can be reasonably assumed that their boxes are too, as is true of so many toys in a series. The box for Olive Oyl depicts her peddling along the road, the tricycle bell ringing, with a nice blue sky in the background. The design shown on the tricycle illustration is basically the same as found on the actual toy except the tires are not striped. The box is marked "Olive Oyl Mechanical Tricycle With Revolving Bell." Probably the only difference in the Popeye box is the substitution of his character and name for that of Olive Oyl.

The toys are hard to find, especially with the boxes. Olive Oyl seems to be more difficult to obtain than Popeye. (Also see Wimpy Mechanical Tricycle.)

Value: $900-$1,000
With box $2,000-$2,500 each

Popeye, with a tin body and celluloid arms and legs, rides a tin tricycle. This wind-up is 4" high and was made in the 1950s. Olive Oyl rides a similar tricycle. Courtesy Smith House, Eliot, Maine.

POPEYE AND OLIVE OYL PLAYING BALL: 1950s. base 19" long, figures 4 1/2" high. Wind-up.

This is an unusual tin toy in which Popeye and Olive Oyl stand at opposite ends of a base. A celluloid ball attached to the center of the base swings back and forth touching the outstretched arms of each figure. An interesting feature of this toy is that Popeye and Olive Oyl each have separate wind-up keys.

As interesting as the toy is, the box is even more so, for it is triangular in shape, and from the side looks rather like a ski slope. The box title is "Popeye and Olive Oyl Playing Ball" and the couple are depicted doing just that. Underneath is the Linemar slogan, the version that resembles the Marx slogan: "One Of The Many Linemar Toys. Do You Have All Of Them?" It is known that the Japanese manufacturer TPS made some boxes like this one, so perhaps it is possible that TPS made the Popeye and Olive Oyl Playing Ball" toy under the Linemar name.

Value: With box $1,500-$1,800

POPEYE AND OLIVE OYL POP UPS WITH SQUEAKERS: 1950s. 7" high each. Push action.

These scarce and curious toys are made of composition and cloth. When the heads are pushed down, their necks shoot up and the toys emit squeaks. Olive Oyl is the more available of the two figures. Her appearance is similar to the way she generally looks except that her dress has a large collar and she carries a cardboard parasol in her right hand. On her chest is a paper label which says "Push Me Down And I'll Pop Up." Popeye's sailor outfit appears the same, but painted on his face are exaggerated lines around his eyes. The toys have Linemar labels on the bottoms of their feet.

The box for Olive Oyl reads "Pop Up Olive Oyl With Squeaker" and shows the smiling Olive Oyl, parasol in hand. Except for his name, Popeye's box heading is the same as that of Olive Oyl. The box illustration shows the sailor man staring at a box of spinach.

Value: For pair with box $1,600-$1,800

POPEYE AND OLIVE OYL SPORTS ROADSTERS: 1950s. 8" long. Friction.

Popeye and Olive Oyl, with large heads, each drive tin convertibles with tin radiators, tin bumpers, tin hubcaps and black rubber tires. Popeye smokes his pipe, but the most striking of the pair is Olive Oyl with a pony-tail so long that her hair flows over the back of the car. Olive Oyl drives either a red or a yellow convertible, but Popeye has only been reported driving a yellow one. However, it is quite possible that Popeye was manufactured driving a red convertible too. The Popeye convertible is harder to find than the Olive Oyl version.

In this 1950s 8" long friction toy, an Olive Oyl with unusually long hair, drives a tin car. Popeye drives a similar car. Courtesy NEW ENGLAND AUCTION GALLERY, W. Peabody, Maine.

The box for the Olive Oyl convertible is red, yellow and green, and reads "Friction Olive Oyl Sports Roadster." Also on the box is "Sculptured Figure With Hair." The box illustration shows Olive Oyl in the yellow roadster and she is depicted with her regular hairdo, not the long hair found on the actual toy. The box for Popeye has not been seen, but it is similar to the box for Olive Oyl.

Value: Excellent $500-$600
Excellent with box $1,300-$1,600

POPEYE AND OLIVE OYL STRETCHY HAND CAR: 1950s. 9 1/2" long. Pull toy.

This unique and infrequently found tin toy consists of Popeye and Olive Oyl on a hand car. What makes the toy so unusual is that both characters have spring-like bodies which expand and contract as they work the handle of the hand car. Besides the springs, the rest of Popeye's and Olive Oyl's bodies consist merely of heads and arms. The hand car handle is striped in a design similar to a barber pole and the hand car's wheels are also striped, but more in the pattern of a kaleidoscope. The toy also came with plain unpatterned wheels.

The exceedingly rare box for the toy depicts Popeye and Olive Oyl, under a spotlight, working the hand car. Popeye's body is expanded, while Olive Oyl's body is contracted.

Value: $900-$1,000
With box $3,000-$3,800

POPEYE AND OLIVE OYL WITH UP AND DOWN HEAD ACTION: 1950s-1960s. 5 3/4" high each. Wind-up.

As both of these tin toys vibrate, their heads bob up and down. Although the toys have the same action, they were manufactured in different poses. Popeye stands with his legs together and his arms crossed over his chest. Olive Oyl, the rarer of the two figures, stands with her feet apart and her hands on her hips.

The toys' boxes accurately depict the actual toys with the minor difference that Olive Oyl has her left arm at her side instead of on her hip. Both boxes are marked with the name of their character plus the words "Wind-up And Down Head Action."

Value: Popeye $700-$1,000
Olive Oyl will be a bit higher than Popeye

POPEYE AND WIMPY WALK-A-WAY TOY: 1960s-1970s. 4 1/2" high. Non-powered.

Finally, a Popeye double-figured toy that - unlike previously listed Popeye toys - is neither rare nor is it priced in a stratospheric range.

The plastic figures of Popeye and Wimpy, joined together and with movable legs, walk down an inclined ramp. The toy will also work on a flat surface if pulled by a weighted string inserted in a loop on Popeye's body. The heads of the characters are mounted on springs and shake as the toy moves.

The box for the toy reads "Popeye And Wimpy/Another Walk-A-Way Toy By Marx/No Winding Required." The smiling characters are depicted walking outdoors, while behind them is a background of foliage.

Marx also made a walking toy of Popeye without Wimpy, which has a somewhat different appearance (see Popeye Walking Toy later in this section).

Value: With box $50-$75

POPEYE BASKET BALL PLAYER: 1950s. 9" high. Wind-up.

In this desirable and fun tin toy, Popeye shoots a basket ball through a net. This action is made possible by having the ball attached to a wire which runs from near the base of the goal post to near the top of the backboard. The wire is fastened to a rod at each end. Popeye propels the ball with his hands, causing it to go through the basket net.

Popeye the Basket Ball Player, manufactured in the 1950s by Linemar, is 9" high. Popeye propels the ball into the basket with his arms. Dr. Malcolm Kates collection, Robert Grubb photo.

Popeye wears his usual sailor suit with a blue cap. On his back are the Linemar logo and copyright symbols for both Linemar and King Features Syndicate. The ball is red, yellow and blue. The base of the goal post is green and the rest is yellow with a white cloth net. There is nice lithography on the backboard showing Olive Oyl and Wimpy who appear to be arguing.

As desirable as this toy is, it is not as difficult to find as some others, but the box is rare. The box illustration shows Popeye successfully scoring a goal while being cheered on by Olive Oyl and Wimpy. The box reads "Popeye The Basket Ball Player" as well as "Linemar Best By Far." An interesting feature of this box is that on the side is the other, less common, Linemar slogan, the version similar to that used by Marx, "One Of The Many Linemar Toys. Do You Have All Of Them?"

Value: $1,000-$1,500

POPEYE CYCLIST: 1950s. 6 3/4" high. Wind-up.

In this toy, as Popeye pedals the high wheel bicycle, the bell rings. The toy is tin except for Popeye's light blue cloth trousers. The same type trousers were used on another Linemar toy, Roller Skater Popeye (see later) and are frequently found in faded condition.

The bicycle is predominantly red and black with lithographed spokes. Olive Oyl's head is lithographed on the bell.

Linemar used the high wheel bicycle with three other riders: Mickey Mouse, Goofy and a cloth-dressed Clown. Back in the 1980s, well-known dealer and auctioneer Lloyd Ralston sold the Popeye version then for five times more than the Clown version.

The box for the toy is marked "Popeye Cyclist" and shows Popeye pedaling the high wheel bike while Olive Oyl and Wimpy look on. The Popeye Cyclist toy should not be confused with the Popeye Mechanical Tricycle which lacks the high wheel and which is much smaller.

Value: $1,200-$1,500

Linemar 1950s Popeye Cyclist rides high wheel bicycle while bell rings. Toy measures 6 3/4" high and is tin except for Popeye's cloth trousers. Courtesy NEW ENGLAND AUCTION GALLERY, W. Peabody, Massachusetts.

Japanese manufactured Clown made with the same bicycle as Popeye. Courtesy NEW ENGLAND AUCTION GALLERY, W. Peabody, Massachusetts.

POPEYE LANTERN: 1950s. 7 1/2" high. Battery operated.

This unusual toy features Popeye in the shape of a lantern. His stomach is clear glass and a bulb inside it lights up. His hands are connected to a wire to form a handle. The pipe he smokes is made of rubber.

The box for the toy is marked "Popeye Lantern" and the toy is depicted accurately with the light shining from Popeye's mid-section. The Popeye Lantern is one of the most reasonably priced Linemar Popeye toys.

Value: With box $400-$500

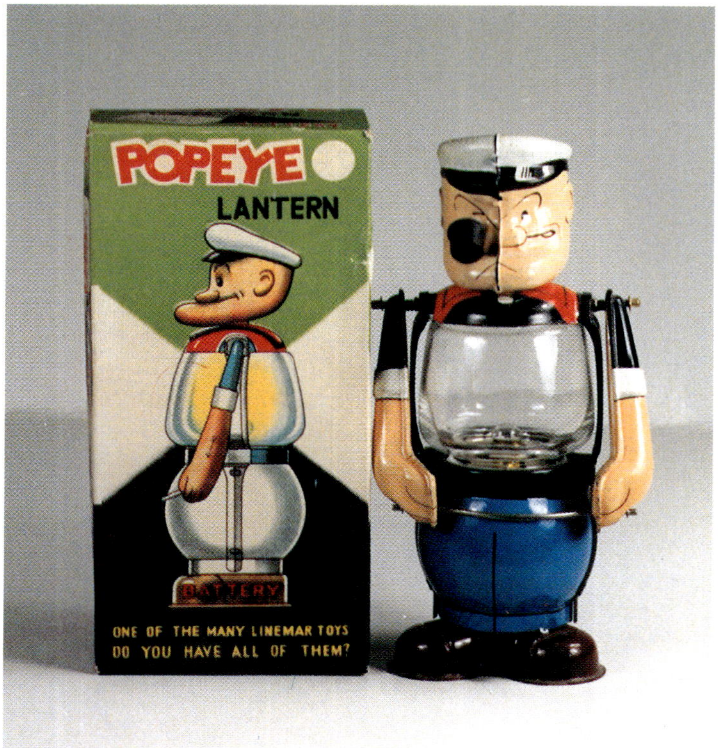

The 1950s Popeye Lantern is 7 1/2" high. A battery powered bulb lights up in Popeye's mid-section. Courtesy SMITH HOUSE, Eliot, Maine.

POPEYE ROWBOAT: 1950s-1960s. 10" long. Battery operated.

Strangely, this remote control tin toy is one of the few, perhaps the only, known example of Popeye in a boat, a theme that would seem to be natural and obvious for a sailor. In the October 1990 *SMITH HOUSE AUCTION CATALOG*, the toy is described as follows: "Extremely rare tin toy has Popeye, with a rubber pipe in his mouth, sitting in a 10" long rowboat with lithographed woodgrained seats/sides/bottom. In action, Popeye moves in rowing action bending at waist which causes the two 6" long tin oars to move as if he was propelling boat. The boat has an adjustable rudder."

All sides of the incredibly rare box show the toy. The front shows Popeye rowing the boat on the water with Olive Oyl and Swee' Pea as passengers. Both passengers look happy, while Popeye is his usual grumpy self. The box title is "Popeye and Row Boat with Moving Oars."

Value: With box $4,000-$4,500

POPEYE TRANSIT CO. MOVING TRUCK: 1950s. 13" long. Friction.

Primarily blue, yellow, cream and red, this Linemar tin truck has characters from the Popeye comic strip lithographed all over it. On the side of the truck, Popeye is illustrated holding a suitcase in his left hand and an open can of spinach in his right hand. Swee' Pea's head is also shown. Between the two characters are the words "Popeye Transit Co./Local Long Distance Moving." On the top of the cab, Olive Oyl's head is depicted and, on the top of the truck, Wimpy and other characters associated with Popeye are seen.

The original box reads "Hauler and Trailer" and accurately depicts the toy.

Value: With box $3,000

POPEYE TURNOVER TANK: 1950s. 4" long. Wind-up.

This tin tank is lithographed all over with characters from the Popeye comic strip. The most available version is predominantly silver, but a rarer version comes in dark blue. The multi-colored lithography is the same for both versions. A flat two-dimensional figure of Popeye lifts the tank from underneath, causing it to turn over.

Popeye is shown on the front of the tank, on the turret and also on the right side casually leaning out of a lithographed "window" with his right hand on his hip as he aims a gun. Wimpy is seen on the left side and Olive Oyl is on the back.

Linemar 1950s tin 4" long tank is turned over by Popeye from underneath. Samuel Caley collection, R. Grubb photo.

On the lithographed hatch cover are the Linemar logo and copyright symbols for both Linemar and King Features Syndicate. Other similar tanks are the Casper Ghost Turnover Tank, the Flintstone Turnover Tank and the Jetson Turnover Tank.

Value: Silver version $350-$400
Blue version $400-$450

POPEYE UNICYCLIST: 1950s-1960s. 6" high. Wind-up.

In one of the less common wheeled toys, Popeye pedals a unicycle. The toy is tin, except for Popeye's cloth blue trousers, which are often in faded condition. The unicycle is red, white and black with tin lithographed tires. A similar toy was made with Mickey Mouse as well.

Value: $700-$800

POPEYE WALKING TOY: 1960s. 3 1/4" high. Non-powered.

This inexpensive plastic toy consists of Popeye with movable legs pushing a wheelbarrow-like vehicle with a movable wheel at the front. Inside the wheelbarrow is a giant can marked "Spinach." The toy will walk down an inclined ramp or across a level table if a weighted string is attached to the front of the toy and dangled over the table edge.

This Popeye has a different appearance than the Popeye and Wimpy Walker described earlier, being somewhat smaller and with his chin jutting out more.

Value: $50-$75

ROLLER SKATER POPEYE: 1950s. 5 3/4" high. Wind-up.

In this fine toy, Popeye uses his right foot to push himself forward as he skates. Attached to his right hand is a tray holding a can of spinach. The toy is tin except for Popeye's wooden pipe and cloth trousers. There is also a Mickey Mouse Roller Skater in this series, but Mickey holds nothing in his hand. Three other similar skaters were manufactured by the Japanese manufacturer, T.P.S. These latter toys consist of a black chef and a white chef, both holding trays of food, and a hobo.

The Roller Skater Popeye toy is not common but it can be found. However, the box is difficult to obtain. The title on the box is "Roller Skater Popeye" and shows him skating along, holding a can of spinach in his hand without the tray, while an admiring Olive Oyl looks on.

Value: With box $2,000-$2,500

This 1950s Linemar wind-up Roller Skater Popeye measures 5 3/4" high. Tin toy has cloth trousers. Courtesy SMITH HOUSE, Eliot, Maine.

Skating Chef was manufactured by the Japanese firm of TPS. It is a close variation of the Linemar Roller Skater Popeye. Courtesy NEW ENGLAND AUCTION GALLERY, W. Peabody, Massachusetts.

The Skating Hobo is another variation by TPS of Roller Skater Popeye. Courtesy NEW ENGLAND AUCTION GALLERY, W. Peabody, Massachusetts.

Battery-operated Linemar tin Smoking Popeye was manufactured in the 1950s and is 9" high. Popeye waves with his right hand, raises a lighted pipe with his left hand and blows smoke from his mouth. Courtesy LLOYD RALSTON GALLERIES, Fairfield, Connecticut.

SMOKING POPEYE: 1950s. 9" high. Battery-powered.

One of the most striking of the Popeye toys, this battery-operated Linemar tin toy features Popeye sitting on top of a can of spinach. Unlike most of the other toys where Popeye wears a blue, red and black sailor suit, here he wears a white sailor outfit with blue trim and oversized brown shoes. Also unusual is Popeye's expression which could pass as a smile instead of his usual grumpy expression.

The large green can on which Popeye sits is marked "Popeye Spinach." Lithographed on the can is a bowl so filled with spinach that the spinach cascades over the edge.

The action of the toy is for Popeye to wave with his right hand, turn his head, raise his lighted pipe with his left hand and blow smoke from his mouth.

The colorful box for the toy is marked "Smoking Popeye" in multi-colored letters. Popeye seems to be indicating his name. He is detailed accurately, a rather large red airplane emphasizing his lighted pipe. Wording reads "See his pipe light up as he smokes." Seated nearby is the baby, Swee' Pea, also in white. In the background is the water with an ocean liner, plus a cloud-filled blue sky.

Value: $650-$1,000
With box $1,500-$2,000

TUMBLING POPEYE: 1950s. 5" high. Wind-up.

In this ingenious tin toy, Popeye's arms twirl, causing him to turn over. The toy has extra long, jointed arms and short legs to facilitate this process. The pipe Popeye smokes is made of wood.

The box for the toy is titled "Tumbling Popeye" in multi-colored letters and shows Popeye getting ready to somersault.

Value: Excellent $750-$1000
Excellent with box $2,000-$2,200

Tin 1950s Popeye is 5" high. His long twirling arms cause him to turn over. Courtesy Smith House Auctions, Eliot, Maine.

WIMPY (see after Wooden Popeye)

WOODEN POPEYE: 1950s. 4" high. Non-powered.

This unusual and interesting Popeye toy is almost completely made of wood. The only metal parts are the stem of Popeye's pipe and the spring on which his bobbing head is mounted.

The hand-painted Popeye wears a black shirt with yellow buttons and a red collar, blue pants and a blue cap that looks like it is about to fall backwards off his head. On the front of each arm is an anchor and on the back of his left arm is stamped the Linemar logo, while on his right arm is "King Features Syndicate."

The primarily yellow box for the toy is titled, appropriately enough, "Wooden Popeye" in black and red letters, and he is illustrated in profile underneath.

The Wooden Popeye is difficult to find, especially with the box.

Value: With box $500-$750

Rare 1950s Wooden Popeye is 5" high. His spring-mounted head shakes. Foot has been mended. Samuel Caley collection, Robert Grubb photo.

WIMPY:

At least two toys of Popeye's friend Wimpy were manufactured by Linemar. They are much harder to find than Popeye toys.

WIMPY MECHANICAL TRICYCLE: 1950s. 4" high. Wind-up.

This toy is a tin Wimpy with celluloid arms and legs who peddles a tin tricycle. Wimpy wears a white shirt, blue tie, black coat, brown pants and a brown hat. The tricycle is blue, red, yellow and white. As the tricycle moves in circles, a revolving bell rings. Popeye and Olive Oyl ride similar tricycles.

The rare original box reads "Wimpy Mechanical Tricycle With Revolving Bell" and also shows the King Syndicate copyright and the Linemar name and logo. The box illustration shows Wimpy riding his tricycle with zig-zag lines indicating the revolving and ringing bell.

Value: Excellent with box $2,500-$3,000

WIMPY NODDER: 1950s-1960s. 5" high. Wind-up.

Manuafactured by Linemar, this tin Wimpy wears a yellow hat, white shirt, blue tie, black coat, blue pants and black shoes. As the figure walks, his head nods. Wimpy has lithographed arms which rest on his stomach. Except for the head, his body is similar to that of Casper the Ghost, but Wimpy is much harder to find.

Value: Excellent $1,000-$1,200

QUICK DRAW McGRAW:

One of the numerous Hanna-Barbera creations, Quick Draw McGraw is a horse of somewhat limited intelligence. His personality was influenced by the character "Clem Kadiddlehopper" created by television comedian Red Skelton. Quick Draw McGraw was first seen on television in 1959 and continued to be seen for several years.

ANIMAL AIRPLANE: Late 1950s-Mid-1960s. 9" long, 9" wingspan. Friction.

This grey, green and red tin airplane manufactured by Linemar is very similar to the Huckleberry Hound Aeroplane (sic) described earlier. The major difference, of course, between the two planes is found in the pilots. In this toy, the vinyl head of Quick Draw McGraw protrudes from the open cockpit.

The plane has the same body and plastic propeller as the Huckleberry Hound Aeroplane, but has different lithography. Unfortunately, it lacks the cartoon characters depicted all over Huckleberry's plane, having instead just the number "538339" on the left wing and the name "Animal Airplane" on the sides.

The box for the Animal Airplane is very close to that of the Huckleberry Hound Aeroplane. The figure of Huckleberry flying the plane is the same, but the illustration of the plane's lithography matches that of the actual Animal Airplane toy. On the side of the box, Quick Draw McGraw, Huckleberry Hound and Yogi Bear are shown.

Value: With box $975-$1,000

QUICK DRAW McDRAW CAR: c.1962. 4 1/2" long. Friction.

Quick Draw McGraw drives a blue tin convertible with rubber wheels. The toy is from the same series as the Huckleberry Hound Car described earlier, plus the Yogi Bear Car described below. Quick Draw McGraw's movable vinyl head is white and yellow and he wears a brown hat. His body is red tin.

The box for the toy is identical to that of the Huckleberry Hound Car and shows Huckleberry Hound driving. However, Quick Draw McGraw's name is on the side of the box.

Value: Excellent with box $100-$150

HOPPING QUICK DRAW McGRAW: 1962. 4" high. Wind-up.

The Hopping Quick Draw McGraw, manufactured by Linemar, is similar to the Hopping Huckleberry Hound described earlier. Like that toy, it has a bent-over, frog-like appearance and big flipper-like feet. The horse wears a lithographed cap on his head, a neckerchief, a star on his chest and a gun and holster at his side.

Value: $150-$175

QUICK DRAW McGRAW VIEW MARX MICRO-VIEWER: 1960s. Approximately 3 1/2" long.

This nice little plastic toy, modestly priced, is shaped like a hotel building. When looking through a window-shaped viewer in the front of the building, while pressing a button on its base, eight different Quick Draw McGraw scenes can be viewed. Micro-Viewers with different shapes were also made for Huckleberry Hound and Yogi Bear.

Value: $12-$25

ROY ROGERS:

The popular cowboy Roy Rogers appeared in films and, in the 1950s, went on television. Those closely associated with him were his wife Dale Evans, friend Pat Brady, horse Trigger and dog Bullet. Besides the toys included in this section, other Roy Rogers toys are a Roy Rogers Rifle and various Roy Rogers Playsets.

NELLYBELLE JEEP: 1950s. 11 1/2" long. Non-powered.

This grey steel Willys jeep of Roy Rogers has some interesting features. The hood can be raised to reveal a plastic engine, while the windshield can be raised and lowered.

On each side of the jeep is a yellow-lettered decal with the name "Nellybelle." The jeep's wheels are black plastic. "Willys" is repeated on both the hood and front bumper.

The jeep came with three yellow-green plastic figures: Dale Evans, Pat Brady and dog Bullet. The Pat Brady figure also came in white while the dog Bullet figure also came in grey and black (see photo).

The toy came in a box marked "Nellybelle Jeep/With Figures Pat Brady, Dale Evans and Bullet." The jeep is illustrated in a somewhat stark style without any details in the background.

Apparently the Nellybelle jeep came in a more deluxe version as evidenced by an advertisement in the 1957 *SEARS CHRISTMAS CATALOG*, p. 245. This ad shows the Nellybelle jeep with five rubber tires (including a spare) instead of four plastic tires. The biggest change is the added plastic figure of Roy Rogers astride his horse Trigger. A separate saddle and bridle were even included. All this sold for $2.83.

A plastic version of the Nellybelle jeep was included as part of a set (see below).

Value: Standard version with box $250-$350

Steel Nellybelle jeep was manufactured in the 1950s and measures 11 1/2" long. A plastic Pat Brady drives with the dog Bullet in back. Ralph Kaufman collection, Dick Kaufman photo.

ROY ROGERS HAULER AND VAN TRAILER: 1954. 15" long. Non-powered.

This toy was shown in the 1954 *SEARS CHRISTMAS CATALOG*, p.218, described as "Colorful lithographed steel van and cab depict famous Roy Rogers traveling troup. Hauler is cab-over-engine type with modern low slung van. Side and rear doors drop open to form ramp. Six vinyl figures include Roy Rogers, Dale Evans, Pat Brady, Bullet and two horses; separate saddles." The toy sold for $1.89.

The vehicle is primarily yellow, blue and red with black and white trim. On each side, Roy Rogers is lithographed mounted on the rearing Trigger. Written in script on the sides is "Roy Rogers, Trigger and Trigger, Jr." Inside an oval shaped ring is "King of the Cowboys" while another ring contains the words "Smartest Horses in the Movies." The top of the vehicle is lithographed with ventilaters and lights. The toy has black plastic wheels.

Value: ?

ROY ROGERS HAULER AND VAN TRAILER WITH NELLYBELLE JEEP: 1956. Hauler and Van Trailer 15" long, Nellybelle Jeep 4" long. Non-powered.

This toy was introduced in the 1956 *SEARS CHRISTMAS CATALOG*, no page number. It is basically the same as the toy described immediately above, but it has a new addition, a red plastic Nellybelle jeep. The Nellybelle jeep that was sold by itself (described earlier in this section) was larger and made of steel. The ad mentions new information that the two horses have three separate sets of saddles and bridles. The toy sold for $1.98.

The box for the set is dark blue. White lettering reads "Roy Rogers Hauler and Van Trailer/Nellybelle Jeep." In smaller writing is "Complete With Authentically Designed Roy Rogers Figures, Horses, Saddles and Bridles."

All Aboard for the Rodeo . . Roy Rogers Hauler, Trailer $1.97

Roy and all the gang are ready for the big rodeo tour. Trigger and Trigger Jr. are waiting to be led up the side and rear gate ramps into the trailer. Roy, Dale Evans and their dog, Bullet, are there along with Pat Brady and his famous jeep "Nellybelle." They've got the riding equipment too. Figures and jeep are 3-dimensional flexible plastic. Sturdy all-steel hauler and trailer is 15½ inches long.

49 N 5557—Shipping weight 2 pounds $1.97

Roy Rogers "Nellybelle" Jeep and Plastic Figure Set $2.83

Dale Evans and Pat Brady take a spin in "Nellybelle," the jeep, while Roy rides Trigger and Roy's dog, Bullet, runs alongside. Steel jeep has movable windshield and hood, dummy engine. 5 rubber tires including spare. 3-dimensional flexible plastic figures. Separate saddle, bridle.

49 N 5584—Jeep 11 inches long. Shipping weight 2 pounds 12 ounces........ $2.83

Manufactured in 1954, the steel Roy Rogers Hauler and Trailer is 15" long. Illustrated in the 1957 Sears Catalog, p. 245.

Value: With box $700-$800

ROY ROGERS HAULER AND VAN TRAILER –BATTERY OPERATED REMOTE CONTROL: c.1957. 13" long. Battery operated.

This version of the Roy Rogers Hauler and Van Trailer, manufactured by Linemar, is the hardest to find. In his useful book, *HAKE'S GUIDE TO TV COLLECTIBLES – AN ILLUSTRATED PRICE GUIDE*, Ted Hake describes the toy as follows: "The toy consists of lithographed metal trailer van and metal cab hauler which has cord to a remote control unit with forward and reverse button ... very colorful lithography is on the van trailer which also has side and rear doors that open and close."[10]

The trailer is similar to the one used in the other toys; however, it is shorter, measuring only 13" long, two inches less than the 15" length of the other trailers. This toy has only three porthole shaped lithographed windows while the other toys have four windows.

The hauler of the battery operated toy varies a bit in appearance from the non-battery toys in that it has no fenders covering the front wheels as the other vehicles do. Then, too, the non-battery vehicles have a window in the cab door and a smaller window in a panel behind the door. The battery version does not have a panel at all.

The box for the toy reads "Roy Rogers Hauler and Van Trailer/Battery Operated Remote Control." The head of Roy Rogers is shown beside the title. Stars dot the background. Also on the box, a pair of hands operate the remote control and the truck itself.

Value: With box $500-$600

ROY ROGERS STAGECOACH WAGON TRAIN: 1950s. 14" long. Wind-up.

This colorful toy consists of a plastic stagecoach, driver and two horses pulling three tin wagons. The plastic stagecoach driver and horses come in white with either grey or red trim. The stagecoach has black wooden wheels.

The first tin wagon is mostly red, blue and dark yellow, with lithographed wheels. "Roy Rogers" appears in script lettering. The second tin wagon is dark green and has a yellow sign which reads "The Home Of Trigger The Wonder Horse/Roy Rogers Ranch." Trigger's head can be seen peeking out behind the sign. This wagon, like the plastic stagecoach, also has black wooden wheels. The last tin wagon, bearing the words "Roy Rogers Ranch," is cream and tan and is lithographed in the style of a prairie wagon. This wagon, like the first one, has lithographed wheels.

The yellow, white and red box for the toy reads "Roy Rogers Stagecoach Wagon Train." "Roy Rogers" is in script and "Stagecoach" and "Wagon Train" are in different styles of print. The wagon is shown rolling along, mountains in the background, while a rider on horseback watches the scene. In a comic strip-style balloon are the words "Watch 'Em Go." The toy moves in a circular pattern.

Value: With box $200-$275

Bottom two rows: 14" long wind-up Roy Rogers Stagecoach Wagon Train, manufactured in the 1950s. The stagecoach is plastic and the three wagons are tin. Top two rows: Davy Crocket Wagon Train is similar to the Roy Rogers Stagecoach Wagon Train. Allan collection.

SUPER HEROES:

With the exception of Batman, the Marx Company manufactured very few of the exciting Super Hero toys. In addition to the toys listed in this section, the company made at least two sets of plastic figures, varying from a few inches to 6 inches high, a picture pistol and a bagatelle game (see later volumes for descriptions).

BATMAN:

The fabulous Batman comic strip/comic book character was created by Bob Kane back in 1939. Batman ran on television with real actors from 1966 to 1968. Then, from 1968 to 1970, the show ran in a cartoon version. Most everyone is familiar with Hollywood's tremendously successful recent Batman films starring Michael Keaton which revived interest in the character and helped to raise toy prices for Batman merchandise.

BATCRAFT: 1966. 6 1/2" high. Battery operated.

In this British-made Marx toy, a plastic three-dimensional Batman, bent over and visible to a few inches below his waist, guides the plastic Batcraft. The figure is blue with a yellow and black bat logo on his chest.

This 6 1/2" high British battery operated toy was made in 1956. Courtesy LLOYD RALSTON GALLERIES, Fairfield, Connecticut.

The strange-looking Batcraft somewhat resembles a small bumper car with a rocket-like projection sticking out behind Batman. The craft is yellow with red trim and has a blue ridge around the base. On the front is a large black and red logo decal marked "Batman." At the top of the craft is a false propeller and below are transparent windows through which can be seen a blinking light. The toy moves erratically when running.

The handsomely illustrated box for the toy shows Batman, illuminated in a flare of light, guiding the craft, while two men (perhaps trying to capture him) run to the scene. The illustration depicts the toy accurately, but there are some color differences, the most noticeable being that Batman's costume is taupe and dark grey, not blue as in the actual toy. The box is titled "Batcraft With Flashing Light." Additional wording is "Battery Operated - Tricky Action" and "National Periodical Publications, Inc. 1966."

Value: With box $400-$500

BATMAN CAR: 1960s. 4 1/2" long. Friction.

A vinyl headed Batman with a tin body drives a tin convertible with rubber wheels. Similar cars are driven by Spiderman, Huckleberry Hound, Quick Draw McGraw and Yogi Bear, Captain America, and Thor.

Value: Excellent $150-$250

BATMAN MARVEL SUPER HERO (see Marvel Super Heroes)

BATMAN MASK SPECS: 1966. 7" long.

Great pair of black plastic eyeglasses shaped like a bat with the wings framing the eyes. The "specs" come on a red, yellow, black and white card illustrated with Batman and Robin wearing the specs. Wording exclaims "Just like your favorite hero!" The Batman logo is shown behind the toy title "Batman Mask Specs." Other wording is "Made By Louis Marx & Co. Ltd. Swansea/[England] Copyright National Periodical Publications, Inc. 1966."

Value: $100-$110

BATMAN WATCH #6801J: 1975. Face 1 5/8" diameter, Strap 7 1/2" long.

This toy watch was introduced in the 1975 *MARX CATALOG*, a late period to bring out a new toy. It was described as "Wind-up ticking sound, actually works. Printed face design has action section and 'glows.' Vinyl strap with metal clasp..." A similar watch with a different catalog designation, #5461J, came without the 'glow' feature.

The face design of the watch shows a full length figure of Batman colored blue, purple and black. The watch is attached to a yellow strap. The toy came on a blister card. Similar watches were made with the characters of Superman and Mickey Mouse.

Value: $150-$200

BAT-MOBILE: 37" long, 15" high, 14" wide. Wind-up spring motor.

It would have been a lucky child who received this toy; however, it is uncertain whether or not this toy was ever produced. The Bat-mobile was shown in the 1967 *MARX CATALOG* featuring TV promotional items. It was described as: "The Only Official Rider Batmobile ... Has push back, wind-up spring motor, steering, hand brake. Of shiny black high impact plastic, includes Bat Motor Noise, play Bat-Phone, Poncho Bat-Cape and Bat gimmicks galore." The Bat-mobile has bat decals on the vehicle sides and fins on the back.

The ad shows both a photo and a drawing of a boy wearing his poncho Bat-Cape, seated in the Bat-mobile. Why Marx found it necessary to have a photo and a drawing duplicating each other is a mystery.

The Bat-Phone, which came with the other Bat paraphernalia, was also sold by itself (see below).

Value: ?

Shown in the 1967 Marx TV promotion catalog, this fabulous plastic toy is a 37" long wind-up. It is uncertain whether or not this toy was actually produced.

BAT-PHONE: 1966. Approximately 8" long. Battery operated.

This red plastic telephone has a decal with the head of Batman on one end of the receiver and the Batman logo (a bat) at the other end. On the front of the phone is a large decal which reads "Bat-Phone." Underneath in smaller letters is "Hot-Line." The telephone "talks," but just what it says is not known. It is conjectured that the phone speaks a number of different phrases relating to Batman. The Dick Tracy Talking Telephone, manufactured around the same time (see earlier in this chapter) says a number of phrases related to Dick Tracy.

It is believed that the Bat-Phone came in two different boxes. The first is plain and marked "1 only #4512 M.O. 'Bat-Phone'." The second is an illustrated display box. One of the illustrations is of Batman and his sidekick Robin. The box has a standing circle at the back, the edge of which is shaped like a bat. Tucked in this shape are the words "Batman" and, underneath, "Hot-Line Bat-Phone."

Value: $250-$550

MINIATURE LICENSE PLATES: 1960s. 4" long x 2 1/4" high.

Miniature tin metal license plates were made for three characters in the Batman series: Batman, the Joker and the Riddler. All were originally packaged in bags with illustrated header cards, but the packages are frequently missing. Miniature license plates were made for certain other Super Heroes such as Spiderman, and for other characters such as those in The Man From U.N.C.L.E. (see earlier in this chapter).

BATMAN MINIATURE LICENSE PLATE:

Green metal license plate shows embossed depiction of Batman in black with red trim. The plate is marked "Batman" in yellow.

Value: $70-$80

JOKER MINIATURE LICENSE PLATE:

Pink metal license plate shows embossed multi-colored depiction of the Joker. The plate is marked "Joker" in yellow.

Value: $50-$60

RIDDLER MINIATURE LICENSE PLATE:

White metal license plate shows embossed multi-colored depiction of the Riddler. The plate is marked "Riddler" in red.

Value: $50-$60

ROLYKINS: 1960s. Approximately 1" high. Ball bearing movement.

The Rolykins are tiny, bell-shaped plastic figures with ball bearings in their bases, which allow them to glide along smoothly. The Marx Company made many Rolykins with Disney and other figures, and two Rolykins from the Batman series: Batman and Robin.

The Batman Rolykin is blue and grey with a Batman logo on his chest. He wears a black mask. The Robin Rolykin is red with a yellow cape. He, too, wears a black mask.

It is admirable that the Marx Compnay went to the trouble of illustrating fully and differently the miniature boxes for the Batman and Robin Rolykins. In the illustrations, the toys look somewhat less strange than the actual toys as they are not as bell-shaped. The box marked "Batman Rolykins" shows the figure standing in front of a large full moon. The "Robin Rolykins" box shows a small full moon in the background near a mountain. Both boxes have the Marx logo.

Value: $15-$25 each

WALKING INFLATABLE BODY BATMAN: 1966. 19" high. Battery operated.

Surely one of the strangest of all Batman toys is this English plaything. As described in the November 1990 issue of *HAKE'S AMERICANA & COLLECTIBLES CATALOG*, the toy has a vinyl inflatable body and plastic shoes, the feet of which are connected by a metal bar which operates them. Batman walks slowly forward by moving his feet back and forth. He wears a grey suit with a black and yellow Batman logo on his chest, a black mask with a blue nose-piece, a blue cape, and blue gloves and shoes. An instruction sheet was included.

The attractive multi-colored box for the toy shows Batman in a red outfit and yellow shoes, suggesting the possibility that the toy may have been made in those colors as well as the blue and grey outfit of the actual toy. In the distance, the Batmobile can be seen against a futuristic skyline. The box title is "Walking Batman With Inflatable Body" shown in a combination of different styles of print and script. Behind the word "Batman" is the Bat logo. In small letters is "Nat. Periodical Publications, Inc. 1966/Made By Louis Marx & Co. Ltd. Swansea [England]."

Value: $200-$400

CAPTAIN AMERICA TOYS (see under Other Super-Heroes)

OTHER SUPER-HEROES:

MARVEL SUPER-HEROES: 1968. 5" high. Wind-up.

These plastic toys are so inexpensively made that the wind-up key is plastic, too. They are modeled in an acceptable manner to their waists, but their legs are too short and stumpy. However, they are examples of a type of toy which Marx rarely made.

Glued to the bases of the toys are blue cardboard discs with the Marx logo, the name of the particular character, "Copyright [symbol] 1968 Marvel Comics Group/All Rights Reserved/ Made In Japan." There are two small holes in the discs through which the works of the toys can be seen. The toys move with a vibrating motion.

The Marvel Super-Heroes offered three heroes: Captain America, Spiderman and Thor. Captain America, dressed in red, white and blue, wears an "A" on the forehead area of his mask and has a star on his chest and another on his shield. Spiderman has a blue and red suit accented with a spider's web design. The blue-eyed Thor wears a blue outfit with black spots and yellow boots which match his long yellow hair. On his head is a white winged hat; he holds a hammer in his right hand. Around each wrist is a red bracelet.

The same box was used for each of the three toys and was illustrated with all three characters so it could be used with any of them. The front of the box is clear cellophane to see which character is enclosed.

Here is a case of the box being better than the toy. While these toys are merely adequate, their multi-colored boxes are beautifully illustrated. The left side shows Spiderman and Thor running, while all three figures are depicted on the back, illuminated by a burst of fiery light. On the right side of the box Captain America appears ready to punch an unseen evil doer. The top and bottom of the box show the heads of the three Super-Heroes.

The box title is "Marvel Super-Heroes." Also on the box are "J-6257," the Marx name and logo, "1968/Marvel Comics Group All Rights Reserved/Made In Japan/One Of The Many Marx Toys, Do You Have All Of Them?" and "Non-toxic."

Value: Near mint $60-$70 each

The 1968 plastic mechanical 5" high Marvel Super-Hero Thor moves with a vibrating motion. Others in the series include Captain America and Spiderman. Samuel Caley collection, Robert Grubb photo.

MARVEL SUPER-HERO TRICYCLES: 4" high. Wind-up.

This toy consists of a rather squat plastic figure with an oversized head, pedaling a bell-ringing metal tricycle that moves in circles. The toy was made in three different versions, each character wearing his standard costume as described in the section immediately above. In all versions, the front of the tricycle is striped and the bell has a circular design.

In the first version, a primarily red, white and blue Captain America rides. In the second version, a mostly blue and red Spiderman appears. A nice touch on this particular toy is that each wheel is lithographed with a spider web. In the last version, a mostly blue, white and yellow Thor pedals.

The title on each toy box is "Marvel Super-Hero Tricycle." The characters are named and illustrated pedaling the tricycles.

Value: Excellent $90-$110 each

MINIATURE LICENSE PLATES: 1967. 4"long x 2 1/4" high.

A license plate was made for both Captain America and Spider-Man. Each is marked on the back with the Marx logo, name, "Made in Japan," the copyright date "1967" and "Marvel Comics Group All Rights Reserved."

Both the lettering "Captain America" and the figure of Captain America are red, white and blue on a green tin plate.

Spider-Man is red, dark blue and black, with his name in yellow letters on a blue tin plate.

Miniature license plates were also made for Batman and various Man From U.N.C.L.E. characters.

Value: $50-$60 each

These 4 x 2 1/4 inch tin embossed Miniature License Plates were made for Batman, Captain America and Man From U.N.C.L.E. characters, too. Samuel Caley collection, Robert Grubb photo.

SPIDER-MAN CAR: 1960s. 4" long. Friction

A vinyl-headed Spider-Man with a tin body drives a friction convertible with rubber wheels. Similar cars are driven by Batman, Quick Draw McGraw and Yogi Bear. This toy was made in the 1960s.

Value: $150-$300

SUPER HERO EXPRESS: 1968. 13" long ?. Wind-up.

Rare tin train is colorfully lithographed with fifteen different characters from Marvel Comics, as follows: Captain America, The Daredevil, the Gladiator, The Green Goblin, Hawkman, The Hulk, The Human Torch, Iron Man, Matthew Murdock, Number One, Red Skull, Spiderman, Submariner, Thor and The Thing.

Value: $350-$400

SUPERMAN:

The "Superman" character, created by Jerry Siegel and Joe Shuster, was introduced to the public in "Action Comics" in 1938. The character appeared on television played by actor George Reeves in the 1950s and, in the 1960s, appeared in television animation. An animated Superman appeared in movie cartoons in 1941 and in 1951 George Reeves appeared in a Superman film, "Superman and the Mole People." In 1978, the first of four Superman films was made starring Christopher Reeve.

The tin 1960s Spider-Man friction car is 4 1/2" long. The toy has a vinyl head. Courtesy LLOYD RALSTON GALLERIES, Fairfield, Connecticut.

Although not as numerous as the Marx Batman toys, the Marx Superman toys, both from 1950 (see Volume I) and after, are the best super hero toys that the company made.

SUPERMAN TANK: 1950s-1960s. 11" long. Battery operated.

This impressive toy is tin except for Superman's cloth cape. The three-dimensional Superman is attached to the front of the tank by metal handles and metal attachments which curve around his legs. The July 1991 *SMITH HOUSE AUCTION CATALOG* describes the action of the toy as follows: "Tank advances with machine noise as Superman tries to hold the tank back while moving his legs. He pushes tank forward and lifts the front of the tank."

This 11" long battery operated Superman Tank was made by Linemar, circa 1950s- 1960s. Superman stops the advancing tank, pushes it forward and lifts the front. Courtesy Smith House, Eliot, Maine.

Superman is red and blue and the tank is dark green with white, grey and yellow trim. A star enclosed in two circles decorates the hatch cover and the sides. Also on the sides are "Army" and "691 831." The tank is covered with lithographed rivets and has lithographed bogies and treads.

The box for the toy is marked "Superman Tank/With Automatic Forward And Backward Action/Lifting Tank Action." Also on the box is the Linemar name, logo and slogan "Linemar Best By Far."

The box illustration shows Superman raising the front of the tank and pushing it through a brick wall, all accomplished with only his left hand.

Value: Near Mint toy with box $1,900-$2,100

SUPERMAN TURNOVER TANK: 1960s. 4" long. Wind-up.

Made by Linemar (Japan), this tin toy is similar to the earlier Marx (USA) Superman Tank made circa 1940 (see Volume I, p.75) although the lithography differs. It is possible that the same dies were used to produce both tanks as it is known that Marx sent dies overseas.

In action, a flat, die-cut Superman appears to lift the tank from underneath and turn it over. The tank is primarily dark green with yellow trim and has lithographed bogies, treads and tools. Superman wears his standard red cape, but his costume is primarily yellow rather than the usual blue.

Value: $400-$500

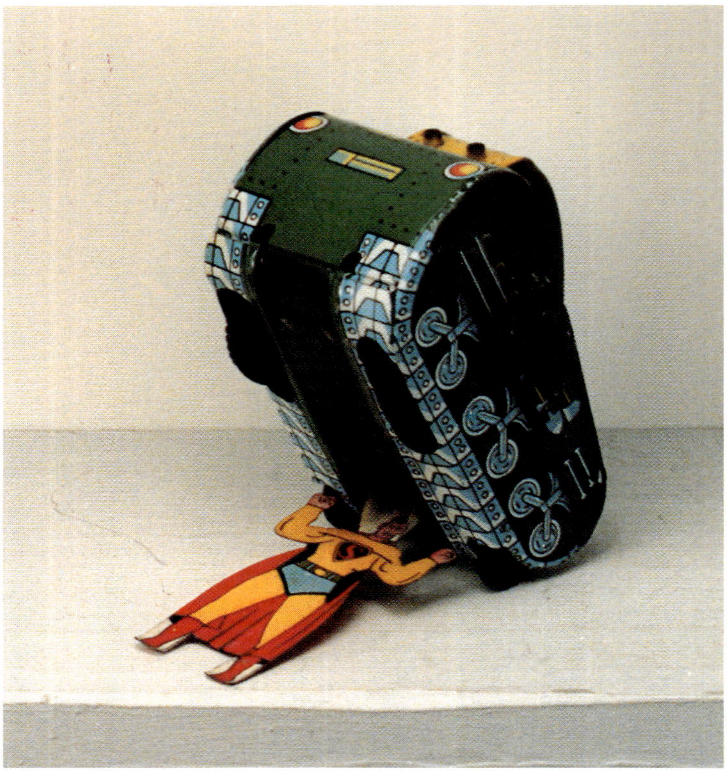

Superman Tank by Linemar. Courtesy SMITH HOUSE, Eliot, Maine.

SUPERMAN WATCH #6801J: 1975. 1 5/8" face diameter, 7 1/2" long strap.

Along with the similar Batman Watch, this toy watch was introduced in the *MARX 1975 CATALOG*. It was described as "Wind-up ticking sound, actually works. Printed face design has action section and 'glows.' Vinyl strap with metal clasp..." A similar watch with a different catalog number, 5461J, came without the 'glow' feature.

The face design of the watch shows a full length figure of a blue and red Superman. The watch is attached to a yellow strap. The toy was packaged on a blister card.

Value: $150-$200

SWORD, SCABBARD AND SHIELD SETS:

The Marx Company manufactured at least five different metal and plastic combination Sword, Scabbard and Shield Sets. In the 1950s, the manufacturer Mattel also made a similar toy, the Prince Valiant Sword and Scabbard.

BEN-HUR SWORD, SCABBARD AND SHIELD SET: 1959. Shields 12" diameter, Swords 18" long.

The best known of the sets, this one was advertised in the 1959 *SEARS CHRISTMAS CATALOG,* the year the film Ben-Hur was made. The set sold for $2.89 and was advertised on the same page (47) as the Marx Ben-Hur Play Set. The Sears ad grandly describes the Sword, Scabbard and Shield Set as being "patterned after weapons used by gladiators two thousand years ago in their battles for wealth and fame in the great arenas of Rome and Antioch." Former Marx employees have stated that the public would have been surprised at the amount of research that went into designing historical-type toys.

The set consists of two metal shields, two metal scabbards and two plastic swords with flexible blades. The heraldic-style lithography on these weapons is outstanding.

Shown in the center of the blue, red, yellow, white, tan and black shield is a detailed depiction of the jeweled handle of the sword. On a banner at the top edge of the shield is the name Ben-Hur. Under this is the Marx logo and riders in horse-pulled chariots. Other illustrations around the shield edge are of a lion, a two-headed bird, a horse's head and a rider on horseback, armed with a sword and shield.

The scabbard has a plastic loop so it can be hung from a belt. It is very ornately lithographed with two heraldic-type lions, several shields, imitation jewels and various decorative motifs. The plastic sword handles are black with three-dimensional imitation red jewels.

The only plain aspect of the set is the unillustrated box, uncolored except for red lettering. Markings are "Ben-Hur Sword, Scabbard and Shield Set" plus the Marx name and logo.

Value: $100-$200

This 1959 Ben-Hur Sword, Scabbard and Shield Set consists of two plastic swords, two metal scabbards and two metal shields. Samuel Caley collection, Robert Grubb photo.

BLACK KNIGHT SWORD AND SCABBARD SET #324: 1975. 18" long sword.

The film "The Black Knight" was made in 1954, however the Black Knight Sword and Scabbard Set was not advertised until its appearance in the 1975 Marx Catalog. Actually, the Black Knight Set was not new in 1975. It was merely a reissued version of the Ben-Hur Set described above. The Black Knigth set came with only one plastic sword and one metal scabbard rather than two swords and two scabbards as in the Ben-Hur Set. Missing entirely are the shields of the other set. The weapons in the Black Knight Sword and Scabbard Set are identical to those in the Ben-Hur Set. It was sold mounted on a card, rather than in a box.

Value: $125-$200

TOM AND JERRY GO CART: 1973. 4" high.

Yet another creation of the prolific team of William Hanna and Joseph Barbera, Tom the Cat and Jerry the Mouse starred as a team in cartoons at M.G.M. Studios in the 1940s. In 1965, cartoons of the pair started in television. Marx made a plastic 1973 4" high friction toy, amusing and reasonably priced. Tom drives the go cart while Jerry sits behind him holding a pistol to his head. The box illustration depicts the toy in a basically accurate manner.

Value: With box $30-$50

TOM AND JERRY RADIO: 1973. 6" long, 1 1/2" wide, 4 1/2" high. Battery powered.

This rather interesting late toy was shown in the 1973 Marx Catalog, p. 38. It was described as being in "Character shape ... plastic, 5 transistor/on/off volume, 9 V btry (sic) supplied. Carry strap, [black] earphones. Boxed. Ages 3 to 9."

The radio is in the shape of Tom the Cat's head which is bright blue. Tom has yellow eyes, a white snout and what is for him an unusually pleasant expression. A brown and yellow Jerry the Mouse leans companionably against Tom's head. The toy has the number 6627HK indicating that it was made in Hong Kong.

TOM AND JERRY SCOOTER: 1971. 4" high. Friction drive.

This is a plastic toy featuring a blue Tom with white and pink trim and a brown Jerry also with white and pink trim. They ride the scooter with Tom driving and Jerry riding in the side car.

The very colorful box realistically depicts the pair driving down a yellow road near bright green grass. The sky above is pink. Another amusing illustration shows an angry Tom running over a banana peel, throwing Jerry up in the air.

Value: With box $30-$50

The 1973 plastic Tom and Jerry Radio is battery powered and is 6" long and 4 1/2" high. Samuel Caley collection, R. Grubb photograph. Value: $50-$75

The 1971 plastic friction Tom and Jerry Scooter is 4" high. S. Caley collection, R. Grubb photograph.

TROMBONE PLAYER: 1950s-1960s. 5 1/2" high. Wind-up.

There is speculation among collectors as to whether this toy resembles Glenn Miller or Tommy Dorsey. Most seem to lean toward the latter. The truth is that although the toy does seem to resemble the band leaders, the resemblance is not close enough to cause any copyright problems for the manufacturer Linemar. In 1935, Marx manufactured the roadside rest gas station with attendants said to resemble Laurel and Hardy. Here too, the resemblance was close, but not close enough to cause trouble.

Also in this series is a figure who resembles band leader Paul Whiteman. He plays a violin just as Whiteman did. The violin player wears striped pants and is thinner than Whiteman, but the resemblance is there. A third unidentifiable figure plays a banjo.

Regardless of all the speculation, the trombone player is an attractive toy. The tin figure is dressed in a band outfit consisting of a red jacket and blue pants. Glasses are lithographed on his nodding head and of course he plays a trombone. His arms move as he plays it.

The name on the box merely reads "Trombone Player" and the trombone player illustrated does not resemble either Glenn Miller or Tommy Dorsey. Following the trombone player are two men playing drums. All wear the red and blue uniform of the actual toy.

Value: Excellent $150-$200

WHITEMAN, PAUL (see Trombone Player)

The 1950s-1960s tin wind-up trombone player is 5 1/2" high. He is thought by some to resemble band leader Tommy Dorsey. Courtesy Smith House, Eliot, Maine.

FOOTNOTES

1. Dates for the television programs and background material for most of the characters in this section have come from the useful book by Ted Hake, *HAKE'S GUIDE TO TV COLLECTIBLES/AN ILLUSTRATED PRICE GUIDE,* Pennsylvania: Wallace-Homestead Book Company, 1990.
2. Gene Scala, "News Article #1," *MARX TOY COLLECTOR NEWSLETTER,* Vol. 2, No. 4, p.9.
3. Correspondence from Roy Ahlgren, February 12, 1992.
4. *HAKE'S AMERICANA & COLLECTIBLES CATALOG #95,* September 1987.
5. Carol Turpen "The Creations of Hanna-Barbera," *ANTIQUE TOY WORLD,* November 1991, Vol. 21, No. 11, p.72.
6. Ibid.
7. Brian Moran, *BATTERY TOYS/THE MODERN AUTOMATA,* Pennsylvania: Schiffer Publishing Ltd, 1984, p.72.
8. Ibid.
9. Richard O'Brien, *COLLECTING TOYS 4TH EDITION,* Alabama: Books Americana, Inc., 1985, pps. 284-285.
10. Ted Hake, *HAKE'S GUIDE TO TV COLLECTIBLES.AN ILLUSTRATED PRICE GUIDE,* Pennsylvania: Wallace-Homestead Book Company, 1990, p. 130.

Disney Toys

INTRODUCTION

It is hard to imagine the modern world without Walt Disney and the impact he has had upon it. Certainly no other animator has influenced the creation of so many toys.

In researching this chapter, the author has found that the big four in Disney "toydom" are Mickey Mouse, Donald Duck, Pluto and, to a lesser extent, Goofy. Unexpectedly, more Donald Duck toys were found than Mickey Mouse toys. Also strange is the overwhelmingly large number of Disney wind-up toys in comparison to Disney battery-operated toys.

Cataloging Disney toys is an extremely challenging task. The various toys are not always as distinct from one another as they might be. To take but one example among many that could be mentioned, consider the "Walking Pluto" toys. There are two plush versions of this toy with Pluto in similar poses, the toy names on the boxes are the same, both toys were made by Linemar, Marx's Japanese subsidiary, and both toys move their heads. The big difference between them is that one toy also moves its tail.

Then there are toys which came in both plastic and metal versions. "Mickey the Musician" made by Marx has a plastic Mickey playing a metal xylophone while the Linemar version "Mickey Mouse with Xylophone" is all metal. It should be mentioned that metal toys frequently came with ears and tails made of some other material, such as vinyl or rubber.

Although bringing order to the bewildering variety of toys was sometimes difficult, it was also enjoyable to study how many truly splendid toys the Marx Company made.

In addition to the toys listed in this chapter, the Marx Company made many other non-moving plastic figures of various sizes that will be covered in a later volume.

Advertisement by the Walt Disney Corporation featuring Mickey Mouse as the world's greatest salesman. The Marx name is shown mid-center on Mickey's right side. Playthings, May 1968, p. 9.

ACROBATS

Marx's Japanese subsidiary, Linemar, manufactured a battery-operated version and also a wind-up version of Donald Duck or Mickey Mouse or Pluto as acrobats. The battery-operated version is described first.

DISNEY ACROBAT: 1950s-1960s. Figure 3 1/2" high, frame 9" high. Battery-operated.

In this toy, a frame holds a trapeze on which Donald or Mickey or Pluto swings. The tin frame is silver. Donald is white with a yellow bill and legs and wears a blue sailor suit with a red bow tie. Mickey is black with a flesh-colored face, red pants and tan shoes. Pluto is yellow. The figures are celluloid, a rather fragile material. Lithographed on the battery box cover are Donald, Mickey and Pluto, swinging on trapezes. At the top of the actual trapeze is a knob which controls the speed at which the toy swings.

The original box is marked "Disney Acrobat" and shows Goofy and Donald, the latter with his cap falling off, swinging from trapezes. On the ground, Mickey rushes to join in the fun. This toy sells for more than the wind-up version as it is rarer. The toy is similar to the wind-up version, but for full comparison, see the description of the wind-up version, "Gym - Toy Acrobat" below.

Made in the 1950s to 1960s, a 5" celluloid Donald Duck swings from the bars of 10" high frame. Courtesy SMITH HOUSE, Eliot, Maine.

GYM-TOYS ACROBAT: 1950s-1960s. Figure 5" high, frame 10" high. Wind-up.

The celluloid figures of Donald or Mickey or Pluto swing from the bars of a tin rust-colored frame. The colors of the figures are the same as in the battery-operated toy except that Pluto is tan instead of yellow. A flagpole at the top of the frame holds a pennant-shaped cream flag. Lettered on it in red is "Gym-Toys." The "Disney Acrobat" previously described is more elaborately made than this "Gym-Toys Acrobat." It has an actual trapeze whereas the "Gym-Toys Acrobat" merely swings on the bars. However, the figures of this toy are bigger, being 5" high compared to the 3 1/2" of the other toy. A further difference is that the other toy does not have the flagpole and flag. The colorful original box is marked "Gym-Toys Acrobat" and depicts Pluto, Donald and Mickey swinging from trapezes. In the distance are circus tents topped with pennant-shaped flags.

Value: Excellent with box $500-$600

The same toy with Mickey Mouse. Courtesy SMITH HOUSE, Eliot, Maine.

The box for the toy. Courtesy SMITH HOUSE, Eliot, Maine.

DONALD DUCK STORE DISPLAY: (See Donald Duck section.)

BABES IN TOYLAND SOLDIER: 1961. 6 1/2" high. Wind-up.

This marching soldier toy comes from the 1961 Disney film, "Babes in Toyland." The figure is tin with a rubber nose. The soldier's red uniform has yellow frog closings and a white crossbelt. On his cap is a yellow material tassel which is frequently missing. His big black mustache matches his black eyes, the brim of his cap, and his belt and boots, while the red and white drum he beats matches his uniform.

This uncommon toy, even rarer with the box, beats his drum as he marches forward. The flat feet, with two finger-like projections on each side, are typical of the feet found on many Linemar robots. The Linemar toy "Walking Pinocchio" also has the same type of feet and also has a rubber nose.

The original box is marked "Mechanical Marching Soldier from Walt Disney's BABES IN TOYLAND." Other markings are the Linemar name and logo. The box is unusually attractive, having great color and graphics. The soldier is shown in the foreground beating his drum. In the background is a tan castle and, in the blue sky, a red and white striped balloon. The stripes are repeated on the sides of the box.

Value: Excellent with box $350-$450

The rare 1961 wind-up Babes in Toyland soldier is 6 1/2" high. Courtesy NEW ENGLAND AUCTION GALLERY, W. Peabody, Maine.

WALT DISNEY'S GO-MOBILE AND SNOWDOMES: (See later in this chapter.)

BAMBI: 1950s-1960s. 6" high. Squeeze action.

This toy, of course, represents the title character from the 1942 Disney film *Bambi.* To operate the toy, one squeezes the handle of the remote control cable attached to the back of the toy and the toy deer hops and makes a noise. Donald Duck and Jiminy Cricket were made in a similar manner.

Bambi has a tan and brown body with white spots and a white chest. The toy has somewhat of a snubbed, round snout compared to the pointed snout of the Disney movie Bambi, but resembles him otherwise. TN of Japan made a toy that is somewhat similar in appearance, but is a wind-up version.

The original box is marked "Jumpy Bambi." It has the Walt Disney copyright symbol "Walt Disney Productions" and the Linemar logo. The drawing on the box looks exactly like the Disney film Bambi.

Value:Excellent with box $375-$400

FLOWER THE SKUNK, THUMPER THE RABBIT: (See these additional characters from the "Bambi" film later in this chapter.)

BIG BAD WOLF AND THE THREE LITTLE PIGS: 1960s. Each 4 1/4" high. Wind-up.

These tin hopper toys are based on the 1933 Disney cartoon of the same name. All three pigs are pink with rather red, round cheeks. One pig wears a red shirt with a blue bow and a yellow cap. In his lithographed right hand is a lithographed flute. Another pig wears a green shirt with a white collar and a red bow plus a green cap. He holds a lithographed violin. The remaining pig wears blue overalls, a yellow shirt and a blue cap. This pig carries a trowel. The black wolf has a flesh-colored snout, pink pants and a pink hat with a blue band.

It is extremely difficult to obtain a complete set of these figures, let alone their boxes. Sometimes individual pigs can be found, but the wolf is rarely seen. Linemar made many hoppers similar to this set, such as members of the Flintstone and Jetson families from the television series.

The original boxes for the individual figures in the set are unique in that each shows story scenes and can be made into a house. The title of the toy, "Big Bad Wolf And The Three Little Pigs" is shown on one of the top flaps of each box. The other two box flaps form the roof.

One side of each box reads, "I built my house of Straw" and shows a scene of the wolf blowing away the straw house and one of the pigs. Another scene on the house is lettered, "I built my house of wood." This time, one of the pigs just runs away. The final scene states, "I built my house of bricks and my friends and I are safe inside." The scene shows the wolf angrily looking in the window of the brick house at the pigs cavorting inside.

Value: Excellent for toys with boxes $2,000-$2,200

Rare 1960s Big Bad Wolf And The Three Little Pigs manufactured by Linemar measure 4 1/4" high each, are tin and hop. The boxes for the individual figures in the set can be made into houses. Courtesy NEW ENGLAND AUCTION GALLERY, W. Peabody, Massachusetts.

CASEY JR. DISNEYLAND EXPRESS: 1955. 12" long. Wind-up.

This floor train consists of an articulated plastic locomotive and three tin cars. The multi-colored train has different Disney characters lithographed on each side of the cars. The first car is lithographed on the left side with the three Donald Duck nephews, Huey, Dewey and Louie, playing "Indians." The other side of this car shows Mickey and Goofy, plus Donald with a horn. The left side of the second car shows Mickey and Minnie Mouse, their nephews and Pluto. On the right side are Dumbo the Elephant and Timothy Mouse. The last car is lithographed on the left side with Pinocchio, Bambi and Joe Carioca - an odd combination. The right side of the last car shows the Wolf and the Three Little Pigs. The train is lettered on each side "Casey Jr. Disneyland Express." The train runs in a zig-zag style. As described in the *SMITH HOUSE TOYS AUCTION CATALOG*, Eliot, Maine, the train: "... puffs along floor in crazy zig-zag pattern. Each car wobbles in different directions as the toy express speeds along its independent way." It is similar to the Davy Crockett Wagon Train (see below).

The original box bears the name "Casey Jr." and a signpost shows the rest of the title "Disneyland Express." Mickey, Joe Carioca, Pinocchio and Donald watch as the train chugs around the bend. Even though the toy was sold in 1955, there is something charmingly old-fashioned in the look of the illustration and titles on this box. It puts one in mind of similar-appearing 1930s boxes. The toy was introduced in the 1953 *SEARS CHRISTMAS CATALOG,* (p.229), selling for 98 cents.

Value: Excellent with box $175-$200

Made in 1953, the 12" long articulated Casey Jr. Disneyland Express has a plastic locomotive and three tin cars. Courtesy SMITH HOUSE TOYS, Eliot, Maine.

CINDERELLA SAND PAIL: Date unknown. 4" high.

As described in *HAKE'S AMERICANA & COLLECTIBLES CATALOG #100,* York, PA for September 1988, this plastic pail is dark blue with a red lithographed tin handle that has small pictures of Mickey, Minnie, Pluto, Goofy and Donald. An oval decal on the pail side shows the head of Cinderella and some mice. Cinderella's name is marked above her head. The bottom of the pail has an incised head of Donald Duck. Under the bottom is a Marx Toys copyright.

It is unusual to find a Marx Cinderella item besides the small plastic Disneykin figures.

Value: Excellent $25-$35.

DANCING DISNEY TOYS: Late 1960s. Each 4 1/2" high. Wind-up.

This plastic wind-up toy was made in several versions including Mickey, Donald, Goofy and Pluto.

Mickey has a black body with a flesh-colored face, wears a yellow shirt and red pants. Donald is white, wears a blue shirt with a red bow and a white cap, and Goofy is black with a flesh-colored snout, wears a red shirt and hat and blue pants. Pluto is yellow with black ears. Attached to the base of the toys is a blue cardboard piece with the Marx name and logo, "Walt Disney Productions" and "Made in Japan."

Despite the "Disney Dancing Toys" name, these toys merely vibrate. They are so inexpensively made that even the

The 1960s vinyl wind-up Disney Dancing Toy Mickey Mouse actually vibrates. This toy was made in the form of other Disney characters as well. Samuel Caley collection, Robert Grubb photograph.

wind-up keys are plastic. The style of the toy and design of the box bear a striking similarity to the Marvel Super-Heroes (see Comic, Film & Television chapter).

The original full-color box reads "Disney Dancing Toys" on all sides. The top and bottom flaps bear the number "J-6191." The toy can be seen through the window in the front and the other sides are illustrated with Mickey, Donald, Goofy and Pluto dancing.

DAVY CROCKETT STAGECOACH AND FRONTIER WAGON: 1950s. Each 5" long. Friction.

Two different tin horse and wagon toys came in the same box: one a stagecoach and the other a frontier wagon, each pulled by a pair of horses.

As described in *NEW ENGLAND AUCTION GALLERY CATALOG*, West Peabody, MA, September 1990, the Davy Crockett Stagecoach is tan with a lithographed passenger in each side window. Lithographed baggage is on top of the stagecoach and in front of the baggage is a black chest marked "Gold" in yellow letters. Marked on the side of the coach is "Davy Crockett Stagecoach."

Lithographed to resemble a prairie schooner, the brown Davy Crockett Frontier Wagon has a yellow tin cover (top) that looks like canvas. Marked in red on the cover's side is "Davey Crockett Frontier Wagon."

The drivers of both vehicles wear tan clothing and the two sets of horses pulling each vehicle are brown. The wheels of the toys are black with open yellow spokes, while the wheels on the horses are plain white.

Once in a long while, these toys turn up separately at shows, but the box shows they were sold together. The original box is marked "Davy Crockett Stagecoach and Frontier Wagon" along with the Linemar name and logo. The box illustration shows the horses briskly pulling the wagons through the desert.

Value: $150-$200

DAVY CROCKETT WAGON TRAIN: 1950s. 12" long. Wind-up.

This floor toy (possibly articulated) consists of two gold-bridled black plastic horses pulling a brown plastic wagon and three tin wagons.

The first car is red, white and blue and is marked "Davy Crockett." It has a portrait of him clad in buckskin. The second car is yellow, tan, orange and black and is marked "Gen'l. (sic) Jackson's Volunteers" and shows them illustrated. The third car is tan and white and is designed like a prairie schooner with markings indicating "Davy Crockett King of the Wild Frontier." (It is similar to the third car of the Roy Rogers Stagecoach Wagon Train.)

This toy is far more common than the Davy Crockett Frontier Wagon and Stagecoach toy described immediately above, but not as common as the similar Casey Jr. Disneyland Express, also described above.

The original box title is "Davy Crockett Wagon Train." Also on the box is "Walt Disney" plus the Marx logo, name and address. The box illustration shows the train being attacked by Indians. On the left side of the box is a portrait of Davy Crockett in a wooden frame.

Value: $150-$200

Davey Crocket Wagon Train. This plastic and tin 1950s toy measures 12" long. Courtesy Trip Riley Collection.

DISNEY AIRPLANE: 1950s. 7 1/2" wingspan. Friction.

This desirable tin airplane was manufactured by Linemar. It is silver with red trim and has four turning propellers. In this handsome toy, the fuselage and wings are lithographed with the characters Mickey, Donald, Goofy and Pluto. The wheels are tin.

Value: With box $1,000-$1,200

The 1950s tin Disney Airplane was manufactured by Linemar. The friction toy has a 7 1/2" wingspan. Courtesy SMITH HOUSE, Eliot, Maine.

DISNEY CARS: 1950s-1960s. Each 5" long. Friction.

In this toy, there are three separate cars driven respectively by Donald Duck, Goofy and Mickey Mouse. The figures and cars are plastic except for the black rubber wheels. All of the open cars are red with black and gold trim plus running boards. Mickey's and Goofy's cars also have blue and orange trim.

Although the cars are similar in appearance, they differ in certain details, such as Goofy's car having a back seat as well as a front seat while the other cars have only front seats. Additionally, Mickey's car seat is larger and straighter than Donald's car seat.

These cars came with paper labels on the base which have the Marx logo, Disney copyright and "Hong Kong Walt Disney Productions." Embossed on the base of each car is its type: Donald drives a Stutz Bearcat, Goofy drives a 1910 Ford and Mickey drives a Stanley Steamer.

Donald Duck wears his blue sailor suit and cap, Goofy is dressed in a red sweater and hat plus blue pants and Mickey wears his familiar red pants and a blue hat.

Value: $75-$125 each.

Three Marx Hong Kong friction cars of the 1950s-1960s. The 5" long toys are plastic with rubber wheels. Dr. Malcolm Kates collection, Robert Grubb photograph.

DISNEYLAND JEEP: 1950s-1960s. 10" long. Friction.

This tin jeep is blue with a folding windshield. It is lithographed all over with various Disney figures, including Mickey on the hood and Donald and Jiminy Cricket on the sides. Over the back wheel in multi-colored letters is "Disneyland Jeep." Jiminy Cricket points to the name. The wheels are black rubber.

The original box is titled "Disneyland Jeep." Mickey and Goofy are depicted looking at the jeep.

Value: $200-$300

The 1950s-1960s friction-powered Disneyland Jeep is tin with rubber tires and measures 10" long. Ed Owens collection.

DISNEY KART: 1960s. 6" long. Friction.

This tin cart is driven by either Mickey, Donald or Pluto. It is tin with black rubber wheels and the characters' heads are vinyl.

Made by Marx, Japan, the toy has rather high handlebars. It should not be confused with the "Mickey Mouse Racing Kart" which is a different toy (see Mickey Mouse toys section).

Value: Excellent $250-$300

The 1960s Marx Japan Disney Kart measures 6" long and is friction-powered. The Kart is made of tin and Pluto's head is made of vinyl. Courtesy LLOYD RALSTON GALLERIES, Fairfield, Connecticut.

DISNEYLAND LICENSE PLATES: 1960s. Each 2 1/4" high x 4" long.

These three tin miniature license plates each bear a different Disney character. Each plate is designed with the embossed head and first name of the character. "Mickey" is in red on a yellow background. "Minnie" is in yellow on a pink background. "Snow White" is in white on a blue background. The rest of the colors are not known, but each is different.

Marx made Miniature License Plates in several other versions, including those for characters from "The Man From U.N.C.L.E." and certain Super Heroes including "Spiderman."

The license plates were generally sold in bags with header card titles.

Value: Excellent each $50-$75

DISNEY SET (4 piece): Early 1950s. About 4" long. Friction.

This yellow plastic toy consists of a four piece character set on black rubber wheels. This is a curious set in which Donald flies an airplane, Mickey drives a tool truck and Goofy drives a car. Strangely Pluto, lacking a vehicle, just lies prone with a plastic ball between his paws. The same Mickey and Pluto figures were used in a later 1954 set with one other toy (see below).

Value: Excellent with box $100-$200

DISNEY SET (3 piece): 1954. About 4" long. Friction.

The plastic Disney three-piece character set was introduced in the 1954 *SEARS CHRISTMAS CATALOG* where it sold for $1.19. The ad stated, "New Disney friction toys ... amusing Disney toys provide lots of fun and laughter as they run forwards or backwards."

This set is painted in multi-colors instead of just yellow like the earlier 1950s set, but also has black rubber wheels. It retains Mickey in the tool truck and Pluto with the ball, but deletes the Goofy car and Donald airplane. The deleted figures are replaced with Donald driving a tractor. These toys sometimes turn up separately at toy shows.

Value: Excellent $150-$200

DISNEY SPEED BUGGIES #6438HK: 1974. 6" long. Ratchet/fly wheel.

This plastic toy consists of a three-car set, with each car driven by a different Disney character. Motion is provided by a plastic ratchet pull cord which operates a fly wheel.

The three cars are driven respectively by Mickey, Donald and Pluto, whose heads protrude from the drivers' seats. The black fly wheel can be seen in the middle of each car. However, each car is different. Mickey's car is pink, Goofy's car is wine-colored and Donald's car is yellow and red. Mickey's and Goofy's names are marked on the car hoods and Donald's name is marked on the side of his car. Mickey's car has both front and rear fenders, Donald's car has only rear fenders and Goofy's car has neither front nor rear fenders. All cars have black and white plastic wheels. (Description from the 1974 *MARX CATALOG.*) Note that the letters "HK" in the catalog number indicate that the toy was made in Marx's Hong Kong branch.

Value: ? ?

DISNEY VAN TRAILER SET: 1963. Each 12 1/2" long. Friction.

This uncommon set actually consists of three toys, but fuller details are available for only two of the three van trailers. Manufactured by Linemar, each of these three toys has a cab pulling a steel trailer which is lithographed with a different multi-color Disney character design. As described in *PLAYTHINGS,* April 1963 (p.115), all have nickel-plated bumpers, windshield trim, hubcaps and an opening double-door in the rear. Each toy sold for 79 cents![1]

DISNEYLAND VAN TRAILER depicts Goofy, Minnie and the nephews of Mickey Mouse and Donald Duck riding a train, while Donald, Daisy and Mickey watch and Pluto, Chip and Dale run toward it. The trailer is marked on the front "Walt Disney's Disneyland Co."

Value: Excellent with box $2,000-$2,500

SNOW WHITE VAN TRAILER is lithographed with Snow White and the Seven Dwarfs. It is marked "Snow White" on the front. A blue cab pulls the trailer.

Like so many sets, individual toys from this one turn up at toy shows. The toys are illustrated realistically on the boxes.

Value: Excellent with box $2,000-$2,500

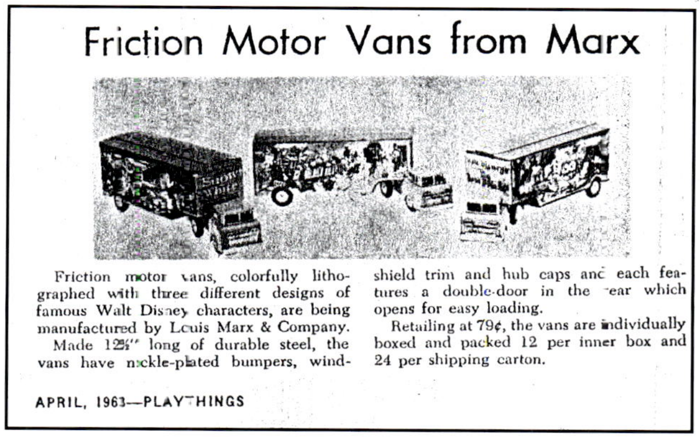
Friction Motor Vans from Marx

Friction motor vans, colorfully lithographed with three different designs of famous Walt Disney characters, are being manufactured by Louis Marx & Company.

Made 12½" long of durable steel, the vans have nickle-plated bumpers, windshield trim and hub caps and each features a double-door in the rear which opens for easy loading.

Retailing at 79¢, the vans are individually boxed and packed 12 per inner box and 24 per shipping carton.

APRIL, 1963—PLAYTHINGS

Friction Motor Vans from Marx advertisement, including the Snow White Van Trailer on the left. Playthings, April 1963, p. 115.

Snow White Van Trailer, made in 1963 is friction and measures 12" long. It is part of a three piece set called Disney Van Trailer Set. Courtesy NEW ENGLAND AUCTION GALLERY, W. Peabody, Massachusetts.

WALT DISNEY CHARACTER CAROUSEL: 1950s. Wind-up.

The toy is 7" high. Its 6" diameter base has a 2" high x 2" diameter round stand. A rod extends from the base with a 6" diameter umbrella. The toy is made of tin with celluloid figures which are 2 1/2" high.

As shown in the *SMITH HOUSE AUCTION CATALOG,* Eliot, Maine, January 1989, the base of the yellow, blue and red toy reads "Walt Disney Character Carousel." The round stand is lithographed with Donald Duck's nephews and the top of the umbrella is lithographed with Mickey, Donald, Goofy and Pluto. Suspended from the umbrella are four tin balls and four swings with the celluloid figures of Mickey, Donald, Goofy and Pluto. When wound, the umbrella turns and the swings rock.

The original box is also titled "Walt Disney Character Carousel" and the head of either Mickey, Donald, Pluto or Goofy, each within a circle, appears at each corner. One each of these characters is depicted full length on each side of the box.

Value: Excellent $1,000-$1,250

WALT DISNEY DELIVERY WAGON: 1950s. 6" long. Friction.

In this toy, either Mickey, Donald or Pluto ride the front part of a tricycle to which a wagon is attached. The tricycle is made of tin while the heads of the characters are made of celluloid.

The white-feathered Donald Duck wears his usual blue sailor suit with a red bow tie and a blue cap. Mickey's face is white rather than the more usual flesh color and he wears red pants and a red hat. Pluto is yellow-tan. The wagon's interior is mustard colored; the wagon's exterior is blue decorated with lithographs of the characters' heads. Marked on the side is "Mickey's Delivery." The tricycle is red. The white wheels of the tricycle have cut-out yellow spokes and are believed to be tin, while the wheels on the wagon are thought to be black rubber.

The original box is marked "Walt Disney Delivery Wagon" and has the Linemar name and logo. The illustration shows Donald prepared to drive the wagon as Mickey climbs in, while Pluto runs alongside.

Value: With box $500-$600

In this 1950s friction toy, Pluto and Donald Duck drive the 6" long wagon. The toy was also driven by Mickey Mouse. Courtesy NEW ENGLAND AUCTION GALLERY, W. Peabody, Massachusetts.

WALT DISNEY GO-MOBILES: Early 1960s. Each 6" long. Friction.

This wheeled vehicle is variously driven by each of the following characters: the Babes in Toyland Soldier, Mickey Mouse, Pinocchio and Professor Von Drake. The steel characters have vinyl heads. Color descriptions are available for two of the four characters.

The Babes in Toyland Soldier wears a red uniform with white and yellow trim and black boots. His Go-Mobile is yellow and red.

Professor Von Drake wears a blue-grey coat, beige waistcoat, white shirt and black tie. On his white face are black glasses which match his black hat and he has thick grey eyebrows and grey hair. His bill and legs are yellow. Small color details can vary slightly. For instance, this toy has been seen with either white or blue eyes, both colors with black pupils. The Go-Mobile is blue and red. Embossed on the base is the Linemar name and logo, 1961, and the Walt Disney copyright. The wheels of all the Go-Mobiles are black rubber.

The heads of the toys are easily detached and care must be taken when buying one of these toys to be sure that the correct head is on the correct toy.

As described in *PLAYTHINGS,* June 1962 (p.126), each of the toys was sold in a display box marked "Walt Disney Go-Mobile." The side of the box reads "With New Strong Friction Motor." The illustrations show the Go-Mobile being driven by the various characters. The toy has a baked enamel finish, nickel-plated wheel supports, a steering wheel, wheel shaft and hub caps.[2]

Value: Excellent including display box $250-$300

WALT DISNEY MECHANICAL TRICYCLES: 1950s-1960s. Each 3 1/2" long. Wind-up.

This is a Linemar toy in which either Mickey, Donald or Pluto pedals a tin tricycle.

All the characters were made in celluloid and, in addition, Donald and Pluto were made in versions with tin heads and bodies and with celluloid arms and legs.

Mickey is black with red pants and yellow shoes. Pluto is orange with black ears and a red collar. The white-feathered Donald wears his usual blue sailor suit and blue cap. However, rare variations exist in which both the tin and celluloid versions of Donald wear red caps.

The red, yellow and white tin tricycle comes with two different designs. One tricycle has a diamond design on the front and a kaleidoscope-like pattern on the tin wheels. Mickey Mouse rides a tricycle with this pattern.

The other tricycle has a flowery design on the front with squirrels and rabbits on the wheels. Both Pluto versions ride this tricycle. The various Donald Ducks ride tricycles with both designs. It is possible that all the characters were made riding both versions. All tricycles have striped bells and striped wheel rims. On the back of the celluloid Pluto tricycle version, and possibly all the versions, is a license plate with the Linemar logo, the Walt Disney copyright and "LM-257."

The tricycle runs in a circle while a revolving bell rings. The toy is similar to the Popeye and Olive Oyl Mechanical Tricycles.

The original box is titled "Walt Disney Mechanical Tricycle With Revolving Bell" and is illustrated with all three characters riding tricycles and the Linemar name and logo. Also shown is the slogan "Linemar Best By Far."

Value: Excellent with box $600-$700.
Excellent with box for rare variations $1,000 each.

Professor Von Drake and The Babes in Toyland soldier ride Go-Mobiles. The toys are tin except for their vinyl heads. They were manufactured in the early 1960s and are 6" long. Courtesy SMITH HOUSE, Eliot, Maine.

The Linemar Walt Disney Mechanical Tricycle was manufactured circa the 1950s-1960s. A celluloid Mickey and Donald pedal tin bell-ringing tricycles. Courtesy NEW ENGLAND AUCTION GALLERY, W. Peabody, Massachusetts.

The Linemar Walt Disney Mechanical Tricycle was also manufactured with Donald Duck or Pluto, as shown. Dr. Malcolm Kates collection, Robert Grubb photograph.

WALT DISNEY ROCKING CHAIR: 1950s-1960s. Pull toy.

Exceptional and unique, this rare Linemar toy consists of a celluloid Donald Duck seated in a tin rocking chair, the sides of which are lithographed to resemble the elephant Dumbo. The chair rocks when it is pulled at the front by a cord attached to a plastic Pluto.

The white-feathered Donald is dressed in his usual blue sailor suit with a red bow tie and a blue cap. He has a yellow bill and legs. The rocking chair seat is red and the lithographed Dumbo on the sides is grey with a yellow, black-spotted ruffled collar. Pluto is yellow.

The original box reads "Walt Disney's Rocking Chair," "Walt Disney Productions" and "Linemar Best By Far" and has the Linemar logo. The illustration shows the chair being pulled by Pluto with the cord in his mouth, when actually the cord is down by his legs. (Also see listings under names of individual characters.)

Value: $700-$1000

DONALD DUCK

The character Donald Duck was introduced in the 1934 Disney cartoon "The Wise Little Hen." By 1935, he was a star in the Disney Studios and the cantankerous duck's popularity continues today.

Donald Duck is white with a yellow bill and legs. He wears a blue sailor top with a red bow tie. His usually blue cap is sometimes found in red.

CLIMBING DONALD DUCK ON HIS FRICTION FIRE ENGINE: 1950s. Truck is 18" long, Donald is 5 1/2" high. Combination battery and friction.

This spectacular toy consists of Donald climbing up and down a ladder on a tin fire engine driven by Pluto while Mickey stands at the rear. In addition to all the above, the fire truck has a siren. Donald climbs up and down the ladder by battery-action. The truck is friction-powered.

The bright red fire engine is lithographed with Mickey, Pluto and Dumbo, while a three-dimensional red-fire-helmeted Pluto drives the truck. The three-dimensional climbing Donald wears the same helmet and has matching red stripes on his sleeves, and the final three-dimensional figure, Mickey, wears no helmet as he stands at the rear of the truck.

The number "6" is on the door and "F. D. No. 6" and "Disneyland F. D." are on the sides of the truck. The Linemar logo is behind the left rear wheel. The tires are either black rubber or plastic.

In the 1960s, Modern Toys of Japan manufactured a 16" long battery-operated fire engine with a plastic Mickey driving and Donald at the rear.

For the Linemar toy being described here, black letters on the very rare original box read "Climbing Donald Duck On His Friction Fire Engine." Additional white letters read "Battery Operated" and "With Siren." Also in black is the Linemar name, logo and slogan "Linemar Best By Far." Green letters state that Donald "Climbs Up And Down The Ladder Automatically."

The illustration on the primarily yellow box depicts Mickey, Dumbo and Pluto beside the fire engine. Curiously, Pluto is also shown driving the vehicle. Donald stands near the ladder.

Value: Excellent with box $1,500-$1,700

Made in the 1950s, this outstanding Linemar toy consists of a battery-operated Donald Duck climbing up and down the ladder of an 18" long friction fire engine. Courtesy SMITH HOUSE, Eliot, Maine.

DISNEY FLIVVER: 1950s. 5 1/2" long. Friction.

Donald Duck drives this tin cream-colored sportscar which is lithographed in multi-colors with Pinocchio on the hood, Mickey on the trunk, Dopey and one of the Donald Duck nephews and Thumper and Pluto on the sides. Also shown are assorted birds and butterflies. Black lettering on the trunk states "Disney Flivver" and on the right side is the Disney copyright and "Walt Disney Productions" plus the Linemar logo and "Japan." The wheels are black tin.

Donald Duck's head has been seen in three versions: tin, celluloid and rubber. In both the tin and celluloid versions, Donald has a rather long neck and wears a red cap instead of his more usual blue cap. In the rubber version (see photo) his neck is normal sized and he wears a blue cap.

However, in the version seen, the neck might not be original. In the tin and celluloid head version, the lithography on the cars is the same, but the car colors differ. The tin head version has a bright yellow car and the celluloid head version has a red car. All versions work in the same way, by pushing down Donald's head to start the car.

Value: Excellent with box $700-$800

The 5 1/2" long friction Linemar Disney Flivver was made in the 1950s. Donald's head comes in tin, celluloid or rubber, as shown. Tin windshield missing from example. Neck may be replaced. Samuel Caley collection, Robert Grubb photograph.

Variation of the Disney Flivver. Courtesy NEW ENGLAND AUCTION GALLERY, W. Peabody, Massachusetts.

A third variation of the Disney Flivver was also produced (see title page of this book). Courtesy NEW ENGLAND AUCTION GALLERY, W. Peabody, Massachusetts.

WALT DISNEY MECHANICAL WADDLING DONALD DUCK: 1950s-1960s. 6" high. Wind-up.

In this interesting toy, Donald's head and arms move as he waddles from side to side. The toy is made of tin and Donald wears cloth clothing.

The original box title is "Walt Disney Mechanical Waddling Duck." Also on the box are the Linemar name, logo and slogan "Linemar Best By Far." Donald is depicted leaning to one side in the act of taking a step.

Value: ? ?

DONALD DUCK AIRPLANE: 1950s-1960s. 9" long x 9" wide. Friction.

This rare Linemar toy depicts Donald, seated in an open cockpit, flying an airplane which is lithographed all over with Disney characters. The airplane is tin with one plastic propeller. Donald has a vinyl head.

This plane is the same as that used in the Huckleberry Aeroplane, although, of course, the lithography and Huckleberry's head differ.

Value: With box $975-$1,000

DONALD DUCK AND HIS NEPHEWS: 1950s. 10 1/2" long. Wind-up.

In this plastic toy, Donald pulls along his three nephews who are all attached by wires. Donald is dressed in his normal blue sailor suit and matching cap, nephew Dewey wears a yellow shirt and a yellow cap with a blue brim, nephew Huey is dressed in a red shirt and yellow cap with a blue brim, and nephew Louie wears a blue shirt with a blue cap and, to be different, a red brim. The toy's wheels are black plastic. The figures are molded in a swimming position as if they are gliding over the water.

The toy is not uncommon, but the original box is. The box title is "Donald Duck And His Nephews." Under each nephew, in tiny letters, is his name. The box illustration does not show the toy as it actually is constructed but, instead, shows Donald and the unattached three nephews walking along.

A similar plastic toy was made circa the 1970s by the manufacturer Illco. The Illco toy plays music, but it is not as nicely modeled as the Marx toy.

Value: Excellent with box $350-$400

DONALD DUCK AND NEPHEWS CAR: 1960s. 3 1/4" long. Non-powered.

Donald drives a roadster with a rumble seat in which his three nephews sit. The car is diecast metal and the figures are plastic. The non-powered car is colored mostly red and blue.

Value: Excellent $150-$160

DONALD DUCK AND NEPHEWS WITH RIFLES: 1950s. About 10 1/2" long. Wind-up.

In this rare Linemar tin toy with tin wheels, Donald pulls his three nephews who are connected by wires. All the ducks are white and Donald wears his usual blue sailor outfit with matching blue cap, Huey wears a red shirt, Dewey a yellow shirt and Louie a blue shirt. All the nephews have multi-colored caps.

All the ducks carry rubber rifles which are frequently missing. Since the ducks have lithographed limbs, the rifles are fitted in metal slots near the top of their right shoulders, and apparently it is easy to remove the rifles from these slots. Donald Duck And Nephews With Rifles is similar in appearance to "Donald Duck With Huey And Voice" except that the latter toy only has one nephew. (See below.)

Value: With box $1,400-$1,500

DONALD DUCK BALL: 2" diameter.

As described in *HAKE'S AMERICANA & COLLECTIBLES CATALOG #100,* York, PA, this ball is made of clear rubber and has a painted plastic figure of Donald inside. It is yellow-tone.

Value: Excellent $25-$50

DONALD DUCK CLIMBING FIREMAN: 1950s. Figure 5 1/2" high. Wind-up.

Donald is a fire fighter in this Linemar tin toy and climbs an extension ladder which is inserted in the base of the toy.

Forsaking his usual sailor outfit, Donald wears a smart red fireman's coat and helmet plus black gloves. One might not know the figure is Donald, but the base leaves no doubt. It is marked "Donald Duck Climbing Fireman" and there is a picture of Donald's helmeted head. The multi-colored base depicts a scene of Mickey and Goofy, dressed like Donald, hosing down a burning building. Donald Duck's nephews Huey, Dewey and Louie stand with a blanket, ready to catch Daisy Duck who is jumping out of a window.

Although dressed differently, this Donald Duck is the same Donald used in the "Climbing Donald Duck On His Friction Fire Engine" toy described earlier.

The very attractive and uncommon original box is marked "Donald Duck Climbing Fireman" with the Linemar name, logo, number "J-5137" and the slogan "Linemar Best By Far." The scene depicted shows a determined Donald climbing a ladder which rests against a brick wall. Linemar made a similar toy with a non-character duck called Mike Mallard the Climbing Fireman.

Value: Excellent with box $800-$900

A 5 1/2" high wind-up Donald Duck, clad in a fire fighter's outfit, was manufactured in the 1950s. Courtesy LLOYD RALSTON GALLERIES, Fairfield, Connecticut.

DONALD DUCK DIPSY CAR: Late 1950s. 5 3/4" long. Wind-up.

In this Linemar all-tin toy, Donald's spring-mounted head shakes as he drives an eccentric car lithographed with Disney characters.

This predominantly red and yellow multi-colored car has Goofy and Pluto lithographed on top of the hood. Mickey Mouse is on the left side of the car and Thumper is on the right. On the left side of the trunk is Donald, on the right side is Minnie. Only one of the Chipmunks, Chip, is shown on top of the trunk, while Pluto is pictured on the back. On top of the trunk are the copyright symbol and "Walt Disney Prod." Above this is the Linemar logo and "Linemar Co. Inc." Strangely, the latter designation is not on the Mickey Mouse Dipsy Car, which is otherwise marked the same. The car's right side bears a small Linemar Toys logo. "Mickey Mouse" is marked under the lithographed front grille. The car has tin wheels. The large rear wheels have lithographed white spokes edged with red disks and black and white stripes; the small front wheels are red and white. A one inch steel pin-like shape projects from the car's base, which helps to balance the car when it tips back.

Marx manufactured a toy similar to this Linemar version in the early 1950s. The two cars are identical, but the Marx figure is plastic while the Linemar figure is tin. It is known that Marx sent dies to his foreign plants and this toy is an example of such a practice. The Japanese used the cars and made similar metal copies of the original plastic figures. (See *GREENBERG'S GUIDE TO MARX TOYS,* Volume II (p.132) for information on the Marx Dipsy Cars.)

On the original box for the Linemar version, one has to look closely at the illustration to see slight differences from the Marx version. The Linemar toy has the same title as the Marx version "Disney Dipsy Car" although, of course, it is marked with the Linemar name and logo. Both versions show Donald seated in the car which rears up on its rear wheels. However, in the Linemar depiction, the car casts a shadow on the ground and Donald loses his cap, while the Marx version has a striped background and Donald's cap remains firmly on his head.

Value: Excellent with box $975-$1,000

All tin 5 3/4" long Donald Duck and Mickey Mouse Dipsy Cars were manufactured by Linemar in the late 1950s. The Marx Company also manufactured the same toys earlier, but the Marx figures were plastic instead of the metal versions pictured. Philip Rolin collection, Robert Grubb photograph.

DONALD DUCK DISNEY DELIVERY: 1950s-1960s. 4 1/2" long.

In this plastic toy Donald rides the front half of a cycle attached to a rather small cart. To operate, the toy is pulled back and released and the toy then shoots forward. On the back of the toy is a paper instruction label and the name "Disney Delivery." The wheels are black plastic.

Value: Excellent $50-$100

DONALD DUCK DRUMMERS

Donald Duck Drummers come in two versions: an all tin version and one in which the figure of Donald is plastic, but the drum is metal. The all tin version commands higher prices.

DONALD DUCK DRUMMER: 1950s-1960s. 6" high. Wind-up.

In this Linemar all tin version, the drum has mostly blue, yellow and red sides, with a predominantly silver, red, blue and white top. "Donald Drummer" in red letters is lithographed on the top as is Donald's head. Donald rocks back and forth as he beats his drum.

The original box is marked "Donald Duck Drummer" and has the Walt Disney copyright and the Linemar name and logo. Donald is shown happily beating his drum.

Value: $350-$375
With box $750-$850

This 1950s Linemar tin wind-up Donald Duck Drummer measures 6" high. Donald rocks back and forth as he beats his drum. Courtesy NEW ENGLAND AUCTION GALLERIES, W. Peabody, Massachusetts.

DONALD THE DRUMMER: 1950s. 10" high. Wind-up.

In this Marx version, Donald's body is plastic while the drum and drum sticks are tin. As in the all tin version, Donald beats the drum as he rocks back and forth. The drum varies only slightly in color and lithography from that of the all metal Donald. This drum has mostly red and yellow sides with a red, white and silver top. Also, on the drum top is a picture of Donald's head and "Donald The Drummer" in red letters. This popular Marx toy is almost twice the size of the Linemar tin version, but the tin version is more valuable.

The original box reads "Donald The Drummer" with the Walt Disney copyright and the Marx name and logo. The illustration is very close to that on the Linemar box except that Donald's head is thrown back more and his drum is striped instead of its actual triangular pattern.

Value: $150-$200
Excellent with box $350-$400

DONALD DUCK DUMP TRUCK: 1950s. 5" long. Friction.

This very attractive Linemar tin truck has a mostly red cab and a yellow bed with a tailgate that opens and closes. A large Mickey Mouse is lithographed on the inner base of the bed and various Disney characters are lithographed in smaller sizes around the inner sides. Donald Duck is depicted on the outer sides of the bed along with "Donald Dump Truck."

The cab doors are marked "D D" and "Disneyland" is on the grille. The tailgate is marked "Disneyland Dumping Service."

Value: Excellent $300-$400

DONALD DUCK ELECTRIC SCISSORS: Date ? 6" high. Battery operated.

This interesting tin Linemar toy consists of Donald Duck with scissors projecting from his bill. The result of this construction is that Donald only has the top part of his bill. Donald is perched on top of the scissors handle which has a full length illustration of Donald with a normal bill.

The equally interesting box shows an illustation of a hand holding the scissors. Besides the title in red "Donald Duck Electric Scissors" are the words in black "Completely safe for any child/cuts paper only."

Value: $400-$450

Linemar battery operated tin scissors shaped like Donald Duck. The scissors blades protrude from Donald's mouth. Example shown has Linemar sample tag attached. Courtesy NEW ENGLAND AUCTION GALLERY, W. Peabody, Massachusetts.

DONALD DUCK GASOLINE TRUCK: 1950s-1960s. 12" long. Friction.

This Linemar tin toy consists of a cab pulling a gas tanker. As described in *CHRISTIES COLLECTIBLES,* March 1989, the red cab has a circular Mouseketeer medallion (or decal) on the doors. The tanker is lithographed with Mickey and Donald. Between the two characters is "Gasoline Co."

Value: $500-$700

DONALD DUCK HOPPERS

The following two Linemar tin toys both hop, but they are of different heights and they work by different mechanisms.

DONALD DUCK HOPPER: 1950s-1960s. 6" high. Squeeze action.

Squeezing the handle of the remote control cable attached to Donald's back causes him to hop. Donald's arms are lithographed and his hands are folded across his stomach.

Value: $350-$400

(Left hand toy): Linemar tin Donald measures 6" high and was manufactured circa 1950s-1960s. He hops when the handle of the remote control cable attached to him is squeezed.
(Right hand toy): Marx Parade Roadster. For description, see *Greenberg's Guide to Marx Toys,* Volume 1, p.80. Both toys courtesy of LLOYD RALSTON GALLERIES, Fairfield, Connecticut.

DONALD DUCK HOPPER: 1950s-1960s. 4" high. Wind-up.

Donald, with a wistful expression, hops by wind-up power. Donald's arms are lithographed at his sides. The toy has the typical bent-over appearance and big feet of other hoppers such as the characters from the Flintstones cartoon. However, this Donald Hopper is somewhat uncommon.

Value: ? ?

DONALD DUCK IN HIS CONVERTIBLE: 1950s-1960s. 5" long. Friction.

As described in the *LLOYD RALSTON ANTIQUE TOY AUCTION CATALOG,* August 1982, this is a Linemar toy in which a celluloid Donald figure drives a blue, tin convertible car.

Each side of the car has a stripe (color unknown) which dips to a "V" behind each front wheel. Over the rear wheels is "Roadrunner." Donald's head is shown on the trunk. The car has a siren.

The original box title is "Donald Duck In His Convertible." Other markings are the Linemar name and logo and slogan "Linemar Best By Far." Donald is depicted driving the convertible with no other background but a few lines. This toy should not be confused with the Linemar toy "friction convertible with Donald Duck" which has a different appearance. See below.

Value: With box $600-$650

This 4" high Linemar tin wind-up Donald Duck hops. Courtesy of LLOYD RALSTON GALLERIES, Fairfield, Connecticut.

DONALD DUCK MOTORCYCLE: 1950s. 3 1/2" long. Friction.

In this tin Linemar toy, a rather youthful-appearing Donald drives a motorcycle propelled by friction action.

The Donald Duck Motorcycle comes in two color variations. In the first, Donald wears his blue sailor top and cap, red scarf, yellow pants, and brown gloves and boots. The motorcycle is red.

In the second variation, Donald wears a yellow shirt, blue pants and cap, and tan gloves and boots. This motorcycle is yellow.

Both versions have tin wheels with red lithographed spokes. The wheels of the first version are black and white; those of the second version are grey, black and white. Both versions have a small rear wheel made of black rubber to provide balance for the toy.

Similar motorcycles were made with Mickey and Pluto. For photograph, see Pluto Motorcycle.

Value: Excellent $225-$250

DONALD DUCK ON SURFBOARD: 1950s-1960s. About 5" high. Friction.

This is a rare Linemar toy in which a celluloid Donald rides a tin surfboard. As described by Stephanie and Paul Sadagursky, well-known collectors and dealers, in *ANTIQUE TOY WORLD,* April 1984, volume 14, no. 4, p.2, Donald holds a string in his right hand which is attached to the front two corners of the surfboard.[3]

Value: ? ?

DONALD DUCK ROADSTER: 1950s-1960s. 5" long. Friction.

In this Linemar toy, a celluloid Donald figure drives a tin roadster. The toy is very similar to Donald Duck In His Convertible, described above, in that it is also tin and celluloid, with the same type tin windshield and the same size car. However, the figure of Donald Duck is much smaller in the roadster and this toy is not as available as the convertible version.

Value: $650

DONALD DUCK ROLLER: 1970s. 5" high. Non-powered.

In this Marx Hong Kong plastic toy, Donald Duck rolls along on two wheels, one attached to each side of his circularly-shaped body.

DONALD DUCK SCOOTER: 1960s. 4" long. Friction.

This is an uncommon Marx plastic toy in which Donald rides a brown scooter with black wheels.

The original box title is "Donald Duck Scooter by Marx" and the Marx logo also appears on the box. A happy Donald is depicted riding his scooter through a desert-like area. Hills, bushes and a cactus can be seen in the background.

Value: Excellent with box $150

DONALD DUCK SITTING: 1950s. 3" long. Friction.

In this Linemar tin toy, Donald sits on the ground, his two legs stretched out in front of him in a dog-like fashion.

There are many other Disney toys of this type, including Figaro the Cat, Thumper the Rabbit and Flower the Skunk. If you lust for a Linemar toy but do not wish to pay the high prices many such toys command, consider these small friction toys.

Value: Excellent $100-$110.

DONALD DUCK STORE DEMONSTRATOR: Donald, 4" high, Trapeze, 5 1/2" high, Backboard 9 3/4" high. Electrically operated.

In this Linemar store demonstrator, a celluloid Donald swings on a tin trapeze. The base and backboard are wood.

The trapeze is red and the base and backboard are dark green. At the top of the backboard is a black-lettered paper label reading "Walt Disney's Performing Acrobat," the Linemar logo, "Copyright Walt Disney Productions" and the original price of the toy being demonstrated, "$1.79."

Behind the backboard is the electric motor. In front of it is the trapeze, the top rod of which is joined on the left to another rod. This rod, which enables Donald to perform, connects to the motor by extending through a small circular hole in the backboard. On Donald's right side, a counterweight is attached to the top of the frame. There is a winding mechanism on the toy that does not function because of the demonstrator's electric motor.

The Linemar demonstrator is similar to the Linemar wind-up Gym-Toys Acrobat described earlier. The figures of both toys are roughly the same size, but the demonstrator's frame is considerably smaller, measuring only 5 1/2" high to the Gym-Toys Acrobat's 10" high frame.

Value: Undetermined, but high.

Linemar store demonstrator is electrically operated. The 1950s toy resembles the Linemar Gym-Toys Acrobat, but the 5 1/2" high frame is about half as big. Samuel Caley collection, Robert Grubb photograph.

DONALD DUCK WALKER: 1960s. 3 1/2" high. Wind-up.

This is a modest Marx plastic toy in which Donald walks with movable feet.

Although this toy is a wind-up, Donald's appearance, with his painted arms, very short legs and movable feet, resembles the Marx ramp walkers (see later listing below).

Value: Excellent $25-$50

DONALD DUCK WITH HUEY AND VOICE: 1950s. 7" long. Pull toy.

In this Linemar tin toy, a noise-making Donald, with an opening and closing bill, is trailed by his nephew Huey, who is attached to Donald by a string. When Huey is pulled back, Donald moves forward and Huey eventually catches up with him.

Donald wears his usual blue sailor suit and blue cap. Huey wears a red sweater and a red and white cap. The tin wheels near Donald's stomach are unpainted and the tin wheels on Huey's sides are white.

The Donald Duck in this toy resembles the Donald Duck Squeeze Action Hopper described above.

The original box title is "Donald Duck W/Huey & Voice" (sic). The box illustration depicts a standing Donald and Huey with no trace of wheels or a string anywhere.

Value: Excellent with box $900-$1000

This 1950s Linemar Donald Duck With Huey and Voice measures 7" long. When Huey is pulled back, Donald walks forward while making a noise and opening and shutting his bill. Courtesy SMITH HOUSE, Eliot, Maine.

DONALD DUCK WITH SPRINGY ARMS AND NECK: 1960s. 3" high. Non-powered.

In this Marx toy, Donald's spring-mounted arms and head shake. The toy is plastic except for the steel arms and neck springs.

The box title for the toy is "Donald Duck With Springy Arms & Neck" (sic). The toy was packaged in a box uniquely designed as a simulated ship's bridge. Through the clear cellophane on three sides can be seen the ship's wheel and various nautical instruments.

Value: Excellent with box $25-$50

DONALD DUCK WITH BUTTERFLY ON WHIRLING TAIL: 1950s. 6 1/2" high. Wind-up.

In this interesting Marx plastic toy, a Donald wind-up with vibrating action leans on an umbrella while his butterfly-tipped tail whirls.

Donald's umbrella is yellow and the butterfly on his tail is red. This butterfly is frequently mistaken for a bow, possibly because it is solid red without any pattern on it.

The toy was packaged in two different boxes with similar, but different, names. It is believed that the earlier box was a plain, unillustrated one titled "Whirling Tail Donald Duck." The second box reads "Donald Duck Twirling Tail Toy/By Marx." Donald is illustrated, umbrella in right hand, while wavy lines indicate his vibrating action. It is conjectured that the illustrated box may have replaced the plain box in order to stimulate sales.

Linemar also manufactured two tin Whirling Tail Donald Ducks (see next listings), but the Marx plastic version is more easily found.

Value: Excellent with box $200-$300

In this 1950s 6 1/2" high, plastic wind-up, a vibrating Donald Duck leans on his umbrella, while his tail twirls. Courtesy SMITH HOUSE, Eliot, Maine.

DONALD DUCK WITH UMBRELLA AND WHIRLING TAIL: 1950s. 5" high. Wind-up.

This Linemar tin toy differs from the similarly-named Marx plastic toy (described immediately above) in several ways. Donald's left hand is outstretched as if to shake hands and his right hand holds an umbrella, while his tail whirls and his body vibrates. Donald's figure is tin. His red umbrella is cloth with a tin handle.

Value: $700-$800

DONALD DUCK WITH WHIRLING TAIL: 1950s. 5" high. Wind-up.

This Linemar toy is similar to the one described immediately above; however, in this version Donald holds no umbrella, his left arm is more bent and his thumb does not stick out. Donald is tin except for his tail which, in this variation, is rubber. The yellow box for the toy reads "Donald Duck/Mechanical Whirling Tail" and shows Donald holding a butterfly net.

Value: $500-$600

Linemar tin Donald Duck is 5" high and is a wind-up. He vibrates as his tail twirls. Courtesy NEW ENGLAND AUCTION GALLERY, W. Peabody, Massachusetts.

DONALD DUCK WRIST WATCH (See "Mickey Mouse Wrist Watch for 1973")

DONALD THE DRIVER: 1950s. 6 1/2" long. Wind-up.

In this Linemar tin toy, Donald drives a low slung red, pink, white, blue and black car which is lithographed all over with pictures and names of various Disney characters. Mickey and Minnie appear on the hood and Pluto and Goofy are on the doors. Donald is shown on the trunk. The car's wheels are black rubber.

Marx made a Donald The Driver toy in the 1950s using the same car but the Donald figure was plastic. (See *GREENBERG'S GUIDE TO MARX TOYS,* Volume I, p.80.) In this Linemar toy, the Donald figure is similar, but made of tin. The Linemar version was also made with Mickey Mouse as the driver. Like the Marx version, this toy was designed to turn before it falls off of a table's edge. Occasionally, the car does not swerve in time.

Value: Excellent $350-$400

In this 1950s 6 1/2' long wind-up, Donald drives a car designed not to fall off a table. Pictured with Donald the Driver is Donald the Drummer described previously. Courtesy SMITH HOUSE, Eliot, Maine.

Friction Convertible With Donald Duck Driver measures 5" long. Courtesy NEW ENGLAND AUCTION GALLERY, W. Peabody, Massachusetts.

DONALD THE SKIER: 1950s. About 6" high. Wind-up.

In this uncommon Linemar tin toy, Donald skis with a ski pole in each hand.

A larger plastic version of this toy was made circa 1940s-1950s. (See *GREENBERG'S GUIDE TO MARX TOYS,* Volume I, p.77.)

Value: $650-$700

FIRE CHIEF DONALD DUCK WITH CRAZY ACTION FIRE TRUCK: 1950s. 6" long. Wind-up.

The vehicle in this rare and expensive Linemar tin toy is called a truck, although it more closely resembles an Eccentric Car with fire ladders on each side. The car is lithographed all over with various Disney characters including Mickey Mouse and Pluto. One of the greatest features of the car is a three-dimensional tin figure of one of Donald Duck's nephews as a hood ornament. Such ornaments are unusual for Eccentric Cars. On each side are yellow tin ladders that can be moved to upright positions.

Donald is dressed in yellow with orange gloves and a removable red rubber fire helmet with "8" on the front. Curiously, when this helmet is removed, underneath is a lithographed red and yellow cap resembling that of one of Donald's nephews.

Why was the rubber helmet placed over the lithographed cap? Perhaps the cap was used on the toy in error, without the realization that it belonged to a Donald Duck nephew rather than to Donald. Or else, maybe the character was intended to be one of Donald's nephews, but it was later felt that the Donald Duck character would sell better. Another possibility is that the figure was made prior to the decision to use it in a fire truck toy. One can only conjecture as to why this action was taken.

The tin wheels have red, yellow, white and grey hubcaps. The striped tires are black and yellow. There are two large rear wheels and two small front wheels which allow the car to move erratically.

Although the Marx Company made many tin Eccentric Cars, this tin Linemar Fire Chief Donald Duck more closely resembles the plastic Smoky (sic) Sam, Sheriff Sam and Peter Rabbit versions. However, the tin wheels on this Donald Duck car were used on many Marx tin and plastic Eccentric Cars. (See *GREENBERG'S GUIDE TO MARX TOYS,* Volume II, pps.113-132 for Eccentric Cars.)

The original box has the rather lengthy title "Fire Chief Donald Duck With Crazy Action Fire Truck." An amusing illustration depicts Donald falling out of the rearing vehicle while his helmet flies off and the bell on the hood rings. Actually, there is no bell on the toy, although one of the Marx Eccentric Cars (Smoky Sam The Wild Fireman) has a bell.

Value: Excellent with box $1,500-$1,700

NOTE: For additional Donald Duck toys, see previous listings under "Disney Toys."

FRICTION CONVERTIBLE WITH DONALD DUCK DRIVER: 1950s-1960s. 5" long. Friction.

In this Linemar toy, a celluloid Donald drives a red tin car. The yellow box illustration depicts the toy realistically. Wording reads "Friction Convertible with Donald Duck Driver."

Value: Excellent with box $600-$650

DUMBO CAROUSEL: 1950s-1960s. 6" high, 7" diameter. Wind-up.

This mostly tin toy has three Dumbo elephants with rubber tails on a carousel. As described in Debby and Marty Krim's *NEW ENGLAND AUCTION GALLERY CATALOG,* West Peabody, MA, for August 1993, when one Dumbo hits the seven keys at the base of the carousel, a jingling noise is heard.

The multi-colored umbrella top of the carousel is lithographed with Mickey, Donald and Goofy, and turns in the opposite direction to that in which the elephants travel.

Marx manufactured another carousel of similar dimensions, the Disney Character Carousel. However, unlike the Dumbo toy where the elephants are attached to turnstiles and one strikes keys, in the Disney Character Carousel, Mickey, Donald and Pluto rock on swings.

Value: Excellent $1,000-1,100

Dumbo Carousel. This mostly tin toy, made in the 1950s-1960s, is a 6" high wind-up measuring 7" in diameter. When the keys in the carousel base are hit, a jingling sound is heard. Courtesy NEW ENGLAND AUCTION GALLERY, W. Peabody, Massachusetts.

This uncommon Ferdinand The Bull Linemar tin wind-up, manufactured in the 1950s and 1960s, measures about 5 1/2" long. The toy's tail rotates and his head goes up and down. Courtesy SMITH HOUSE, Eliot, Maine.

FERDINAND THE BULL: 1950s-1960s. About 5 1/2" long. Wind-up.

This uncommon tin Linemar toy has vibrating action. He is similar in shape to the tin Marx 1938 version of Ferdinand and, like that version, also holds a cloth flower in his mouth.

However, the differences are greater than the similarities. The Linemar Ferdinand has a black body with a white chest and forelegs, rather than the brown and tan color of the Marx version, and a vinyl rotating tail which has a different shape from that of the Marx toy. Unlike the earlier toys, the head goes up and down. The wind-up key is under his stomach instead of on his side and, finally, he is believed not to have the bee attached to his side as does the Marx version. The Linemar produced Ferdinand is rarer than its Marx counterpart.

Value: $300-$350

FIGARO THE CAT: 1950s-1960s. 3" long. Friction.

This is the Linemar version of the charming cat from the wonderful 1940 Disney animated feature "Pinocchio." The tin black and white Figaro has yellow eyes with black pupils and the inner parts of his ears are pink.

Figaro sits with his two front legs stretched out in front of him. There are many similar toys of this type, including Donald Duck, Flower the Skunk and Thumper the Rabbit.

Value: Excellent $100-$110

For a photograph, see the group shot with Thumper The Rabbit.

FLOWER THE SKUNK: 1950s-1960s. 3" long. Friction.

This is the Linemar version of the skunk from the outstanding 1942 Disney animated feature "Bambi." Flower is made of tin, is colored black and white and sits in the same position as Figaro the Cat described immediately above – that is, with his front legs stretched out in front of him.

Value: Excellent $100-$110

For a photograph, see the group shot with Thumper The Rabbit.

GOOFY

The lovable dimwit Goofy made his debut in the 1932 Disney animated cartoon "Mickey's Revue." By 1935, Goofy's popularity had climbed to star status. After Donald, Mickey and Pluto, Goofy was the character of whom most toys were made.

Goofy has a flesh-colored face with black ears and a black tail. He is generally dressed in the following lithographed clothes: red shirt and hat, black bolero, blue pants with white cuffs and an orange patch on the left leg, white gloves and brown shoes.

GOOFY: 1950s-1960s. 5 1/2" high. Wind-up.

This is a Linemar toy in which a vibrating Goofy's head moves up and down as his tail whirls. Goofy is made of tin with a rubber tail which tends to become rather worn.

The original box for the toy is merely titled "Goofy" and has the Linemar name, logo and slogan "Linemar Best By Far," as well as the Walt Disney copyright. Goofy is depicted standing on the grass in front of a brick wall, with ungloved hands clasped against his chest and a silly expression on his face, accentuated by his back teeth sticking out of his closed mouth. The box side illustration shows Goofy with his arms by his sides and his mouth open and, except for the ungloved hands, this is a more accurate representation of the actual toy. Movement is suggested by a circle around Goofy's tail and lines near his neck.

Goofy was packaged and sold separately (as indicated here) and was also packaged and sold in a different box together with Playful Pluto (see "Playful Pluto And Goofy" in Pluto section).

Value: Excellent with box $500-$600

The tin Linemar 5 1/2" high Goofy was manufactured in the 1950s-1960s. His head moves up and down as his rubber tail whirls. Courtesy SMITH HOUSE TOYS, Eliot, Maine.

This is a 6 1/2" high Linemar toy manufactured in the 1950s-1960s. A tin Goofy rides his cycle while a bell rings. Courtesy LLOYD RALSTON GALLERIES, Fairfield, Connecticut.

GOOFY CYCLIST: 1950s-1960s. 6 1/2" high. Wind-up.

This is a Linemar toy in which Goofy rides a bell-ringing, high-wheel cycle. The toy is tin except for Goofy's cloth trousers.

As described in the *RICHARD OPFER AUCTIONEERING, INC. ANTIQUE TOY AUCTION CATALOG,* March 1992, the cycle is primarily red. The bell is lithographed, not with an image of Goofy's head but, instead, with the heads of Minnie and Pluto. The red tin wheels have white lithographed spokes and black rims. Another version has Mickey Mouse riding a similar cycle.

The original box reads "Goofy Cyclist" and shows Goofy riding the cycle while on his left, Donald runs after him. On Goofy's right, Mickey rides a similar cycle and Pluto runs ahead of the cycle.

Value: Excellent with box $2,000-$2,500

GOOFY'S STOCK CAR: 1950s. 6" long. Friction.

Made by Linemar, this mostly red and yellow car is lithographed with various Disney characters. Goofy drives the car. He is also lithographed on the hood below the number "33." Donald, Minnie, and Pluto are shown on the fenders. Number "3" is on the door. The car is similar in shape to the Mad Hatter's Sky-View Taxi (see below). On the car's roof are the words "Goofy's Stock Car."

Value: $250-$300

GOOFY TWIRLING TAIL TOY: 1950s-1960s. 8" high. Wind-up.

This humorous Marx plastic toy finds a surprised Goofy twirling his tail with a chipmunk firmly attached to it. The brown chipmunk is sometimes understandably mistaken for a squirrel, but it lacks the squirrel's long bushy tail. The toy is common, but the box is rare.

The original box reads "Goofy Twirling Tail Toy/by Marx." It is difficult not to laugh at the depiction of Goofy's truly shocked surprise as the determined chipmunk takes a good bite. The twirling movement of the chipmunk is suggested by a circle around its waist and Goofy's vibrating movements are shown by lines around his body.

Value: Excellent $100-$110
With box $400-$450

The 8" high Marx plastic wind-up Goofy twirls his tail with chipmunk attached to it. Courtesy SMITH HOUSE TOYS, Eliot, Maine.

Linemar 5" high tin wind-up Goofy was manufactured in the 1950s-1960s. He vibrates as a rubber turtle bites his tail. Courtesy SMITH HOUSE, Eliot, Maine.

The turtle can not be seen in this photo. Courtesy LLOYD RALSTON GALLERIES, Fairfield, Connecticut.

GOOFY UNICYCLIST: 1950s. About 5 1/2" high. Wind-up.

In this Linemar toy, Goofy pedals a unicycle which rolls forward and turns. It is tin except for Goofy's cloth trousers.

The unicycle is mostly red and black with yellow tires and black zig-zag lines suggesting treads. Mickey Mouse rides a similar unicycle.

Value: With box $1,500-$2,000

GOOFY WITH WHEELBARROW: c.1960. 5" high, 5" long. Wind-up.

In this plastic toy, Goofy's legs move as he pushes a wheelbarrow.

As described by Fred and Marilyn Fintel in *YESTERDAY'S TOYS WITH TODAY'S PRICES,* the wheelbarrow is marked "Walt Disney's Wheelbarrow" on the side.[4] This toy should not be confused with the Marx tin "Goofy The Gardener" manufactured in the 1940s. (See *GREENBERG'S GUIDE TO MARX TOYS,* Volume I, p.80.)

Value: $25-$50

GOOFY WITH WHIRLING TAIL AND TURTLE: 1950s-1960s. 5" high. Wind-up.

In this Linemar toy, Goofy is tin with rubber ears. He vibrates as a rubber turtle bites his whirling rubber tail. It is rumored that a version was also made with a ceramic turtle. Unlike the other Goofy toys described previously who all wear long pants, this Goofy wears short pants.

The original box reads "Goofy Mechanical Whirling Tail." In the box illustration, Goofy wears long pants and has a look of shocked surprise as he stares back at the turtle biting him.

Value: Excellent with box $600-$700

NOTE: For other Goofy toys, see previous listings under "Disney Toys."

JIMINY CRICKET: 1950s. 6" high. Squeeze-action and wind-up versions.

The appealing little cricket appears in the 1940 Disney animated feature film "Pinocchio." The toy representing him was made by Linemar in both a squeeze-action version and a wind-up version. To use the squeeze-action version, squeeze the handle of the remote-control cable attached to the back of the toy and Jiminy Cricket hops. He also hops in the wind-up version.

Jiminy Cricket is made of tin and wears a large blue top hat with a yellow stripe, a black coat, a white vest and gloves. His cravat and rubber umbrella are yellow. Jiminy's shoes are yellow and brown. Jiminy Cricket has a flesh-colored smiling face with rather pink cheeks.

The original boxes for both the squeeze-action and wind-up versions are similar. Both read "Jiminy Cricket" and he is shown standing in the same pose. However, the box for the squeeze-action version has more elaborate graphics; the squeeze-action box shows the sun circling the first three letters of Jiminy's name as well as tall buildings in the background. The box for the wind-up version lacks these embellishments.

Value: Excellent with box $1,000-$1,500 (for both versions)

LADY AND THE TRAMP

Lady and the Tramp were the two canine stars of the 1955 Disney animated film of the same name.

LADY: 1950s. 4 1/2" long. Friction.

In this Linemar tin toy, Lady, the Spaniel, is tan with dark brown ears and white chest, snout and paws. She wears a red collar and rolls a red, yellow, black and white ball between her front paws.

The original box reads "Lady" and the dog is shown happily looking up at her name. In the background is a building that resembles a church.

Lady is not a common toy and the box is even harder to find.

Value: Excellent with box $200-$275.

TRAMP: 1950s. 4" high. Friction.

In this rare Linemar tin toy, Tramp, the Terrier, is in a sitting position. He has felt ears. Tramp is brown with a lighter snout.

Value: Excellent $125-$150

LOUIE IN HIS DREAM CAR: 1950s. 4 1/2" long. Friction.

In this uncommon Linemar toy, Louie, one of Donald Duck's nephews (the other two are Huey and Dewey) drives his dream car. The car is tin and Louie is celluloid.

The orange car is marked on the side, in red lettering, "Dream Car Disneyland Roadster." Fins jut out at the car front and it has a siren. Louie is dressed in red with a red and white cap.

We are in no doubt as to which of Donald Duck's three nephews is driving the car, because the original box reads "Louie In His Dream Car." A proud Louie is depicted driving the roadster. There are a few lines near the car to suggest speed.

Value: Excellent $400-$600
Excellent with box $600-$700

Made in the 1950s, this friction toy is 4 1/2" long. A celluloid Louie drives the tin car. Courtesy NEW ENGLAND AUCTION GALLERY, W. Peabody, Massachusetts.

MAD HATTER SKY-VIEW TAXI: 1950s-1960s. 5" long. Friction.

This Linemar tin taxi is primarily red, yellow and black, with chrome grille and bumpers, and with various Disney characters lithographed on it. On the hood is "The Mad Hatter's Sky-View Taxi" with a lithographed picture of the Hatter's head. The trunk has the Walt Disney copyright and also "Walt Disney Productions/Disneyland Taxi Service." "Taxi" is on the red and black doors. Projecting from the vehicle top is a sign "Sky-View."

The undercarriage (or works) of the car is lithographed on the base. The red and white tin wheel hubcaps have lithographed spokes and the black tires are rubber.

Although the Mad Hatter's head appears on the hood, strangely he does not appear through the windshield. He is only seen lithographed on the left side, as if sitting in the driver's seat. Instead, a meter and steering wheel are shown through the windshield. Lithographed as if in the left back seat is a duck, presumably Donald Duck, but he does not wear a cap. Lithographed in the rear window are two of the Seven Dwarfs and judging from their expressions, they appear to be Grumpy and Sleepy. Lithographed as if in the right back seat is another of the Seven Dwarfs, but unrecognizable. Through the right corner of the roof, peeps a rather lively Dopey.

Value: Excellent $200-$225.

Linemar 5" long tin friction Mad Hatter's Sky-View Taxi was manufactured circa the 1950s-1960s. Samuel Caley collection, Robert Grubb photograph.

MARY POPPINS: 1964. About 11" high. Wind-up.

Mary Poppins is the title character from the very successful 1964 Disney film of the same name, which combined live action with animation and special effects.

In this Marx plastic toy, Mary Poppins rotates and whirls in a circle. Mary wears a purplish-brown coat, black boots, and a blue hat and skirt. She holds a flowered carpetbag in her left hand and an open umbrella in her raised right hand. Considering the popularity of the film, it is surprising that the toy is rather rare.

The original box for the toy reads "Mary Poppins/Walt Disney's Whirling Toy by Marx." Mary is depicted, umbrella in hand, descending from a starry sky, shooting stars included, but the sky is a bright daytime blue. However, the magical Mary Poppins does not have to be scientifically accurate and the box is very attractive.

Value: Excellent with box $100-$200

MARX PRODUCING WHIRLING MECHANICAL MARY POPPINS

A whirling, wind-up Mary Poppins that will move around on any flat-top surface is being produced by Louis Marx & Company, 200 Fifth Avenue, New York City.

Made of high-impact plastic and hand-decorated in gay colors, the mechanical Mary Poppins is 9½" high. A permanently attached key turns the clock-type spring motor and whirls a bow attached to the back of her coat.

Retailing at $1, the individually-boxed items are packed 12 per inner carton and four dozen per 17-pound outer carton.

Announcement of the production of the Marx Whirling Mechanical Mary Poppins. Playthings, December 1964, p. 69.

MICKEY MOUSE

Not much introductory material is necessary for Mickey Mouse, surely one of the world's best-known and loved characters. Mickey's first movie appearance was in the 1928 silent cartoon "Plane Crazy." However, Walt Disney decided in the same year to use sound in the cartoon "Steamboat Willie" in which Mickey also appeared and Mickey's popularity soared. After the success of "Steamboat Willie," Disney re-released "Plane Crazy" with sound.

Mickey Mouse toys are numerous, but there are not quite as many as there are for Donald Duck. Mickey has a black body and black ears with a flesh-colored face. He generally wears short red pants, yellow shoes and, sometimes, white gloves, but his clothes can vary in color.

DRUMMING MICKEY MOUSE: 1957. 11" high. Battery-operated.

In this outstanding Linemar toy, Mickey walks and beats his drum while his eyes light up and his head nods. The toy is battery-operated by remote control. Mickey has a tin face, as well as tin legs and feet. His body and ears are plush over a metal frame, while his pants are made of cloth. His drum is tin.

Mickey usually wears red and black plaid pants but has also been seen in plain red pants. He beats a red, yellow, white and blue drum. A nice touch is the illustration of Mickey on the face of the drum which also shows him beating a drum.

The original box reads "Drumming Mickey Mouse." Mickey is depicted vigorously beating his drum, a beam of light shining from his eyes. Within the light is "Lighted Eyes."

The toy was advertised at a price of $2.67 in the 1957 *SEARS CHRISTMAS CATALOG,* p.249.

Value: Excellent $600
Excellent with box $1,200-$1,300

The 1957, 11" high, battery operated Drumming Mickey Mouse beats his drum while his eyes light up and his head nods. Courtesy SMITH HOUSE, Eliot, Maine.

Mickey Mouse Drummer advertisement in the 1957 Sears Christmas Catalog, p. 249.

MICKEY CAR: 1950s-1960s. 4 1/2" long. Friction.

In this attractive Linemar toy, Mickey drives a convertible lithographed all over with numerous Disney figures, such as Mickey himself, Donald, Goofy and Pluto. The car is tin and mostly red. The wheels have tin hubcaps and black rubber tires. Mickey is celluloid.

Value: ? ?

MICKEY MOUSE CYCLIST: 1950s-1960s. 6 1/2" high. Wind-up.

This is a Linemar toy in which Mickey rides a bell-ringing, high-wheel cycle. The toy is all tin except for Mickey's cloth pants.

The cycle is primarily red, the bell is lithographed with the heads of Minnie and Pluto, and the red tin wheels have white lithographed spokes and black rims. Goofy the Cyclist is a similar toy.

The original box reads "Mickey Mouse Cyclist" and shows Mickey riding, while on his left, Donald runs along. Behind Mickey, on his right, Goofy rides a similar cycle while Pluto runs behind.

Value: Excellent with box $2,500-$3,000

MICKEY MOUSE DIPSY CAR: Late 1950s. 5 3/4" long. Wind-up.

In this Linemar tin toy, Mickey's spring-mounted head shakes as he drives an eccentric car lithographed with Disney characters.

The mostly red and yellow multi-colored car has Goofy and Pluto lithographed on top of the hood. Mickey Mouse is shown on the left side of the car and Thumper is shown on the right. On the left side of the trunk is Donald, on the right side, Minnie. Only one of the two chipmunks, Chip, is shown on top of the trunk, while Pluto is pictured on the back. On the top of the trunk are the copyright symbol and "Walt Disney Prod." The car's right side bears a small Linemar logo. "Mickey Mouse" is marked under the lithographed grille.

The large rear tin wheels have lithographed spokes edged with red disks and black and white stripes. The small front tin wheels are red and white. A one inch steel pin projects from the car's base, which helps to balance the car when it tips back in typical eccentric car fashion.

Marx manufactured a toy similar to the Linemar version in the early 1950s. The two cars are identical, but the Marx figure is plastic while the Linemar figure is tin. As indicated previously, Marx sent dies to his foreign plants and this toy is another example of such a practice. The Japanese used the car but made a metal copy of the original plastic figure. (For information on Marx Dipsy Cars, see *GREENBERG'S GUIDE TO MARX TOYS,* Volume II, p.132.)

For photograph, see Donald Duck Dipsy Car earlier in this chapter.

Value: Excellent $500-$550
Excellent with box $975-$1,000

This 1950s wind-up Mickey Mouse Cyclist is 6 1/2" high. The toy is tin with cloth pants. Courtesy NEW ENGLAND AUCTION GALLERY, W. Peabody, Massachusetts.

MICKEY MOUSE/DONALD DUCK HANDCARS

Marx made two versions of this plastic yellow handcar, both with Mickey Mouse pumping one side of the handcar's handle and Donald Duck pumping the other. The difference is in the wheels: one version had flanged wheels, the other version had non-flanged wheels.

MICKEY MOUSE/DONALD DUCK HANDCAR: 1954. About 5" long. Wind-up.

This version has flanged wheels and came with 86 inches of grooved, 2-rail track on which it ran. The toy was new in the 1954 *SEARS CHRISTMAS CATALOG,* p.221, where it was advertised for $2.49.

Value: $800-$1100

Handpainted prototype of the Mickey Mouse/Donald Duck Handcar base with illustrations differing slightly from the production base. Joseph Fanara collection.

MICKEY MOUSE/DONALD DUCK HANDCAR: 1955. Handcar, about 5" long; Base, 21 3/4" long. Wind-up.

The plastic yellow handcar is the same one as in the listing immediately preceding, but in this toy it came with a metal base instead of with track. The wheels are flat, non-flanged metal disks which run in a channel in the base.

The base has a three-dimensional black crossing gate and semaphore signal which "warn" when the handcar goes by. The multi-colored base is beautifully lithographed with various Disney characters shown in a rural setting with a church and other buildings. Donald is shown riding a miniature train and Pluto stands with a happy expression. Around the edge of the base, Mickey and Donald are illustrated on the handcar.

The original box reads "Mickey Mouse & Donald Duck Handcar" in the form of a sign held by Mickey and Donald. Lettering within a circle states "Complete with base, semaphore and crossing gate." Mickey and Donald are shown on the handcar while Donald's three nephews, Pluto and, in the distance, Goofy, look at them.

The toy was new in the 1955 *SEARS CHRISTMAS CATALOG* where it was advertised to sell for $2.79.

Value: With box $600-$700

Frankonia's Walt Disney Melody Train Shows D. Duck, M. Mouse

A Walt Disney Melody Train, showing Donald Duck and Mickey Mouse operating a handcar, is now being marketed by Frankonia Products, Inc., 1107 Broadway, New York City.

The train set (#6026), is offered in two sizes. The hand-car is battery-powered and runs around a multi-colored xylophone track. The firm has applied for a U.S. patent for the musical item.

Easter Lineup: Feb. '35

Alexander Doll's Easter lineup included Lady Lovelace, Molly Cottontail, Dicksie & Ducksie, Smoky Tail, David Quack-a-field and Dotty Dumbunnie.

A similar Disney handcar, Frankonia's Walt Disney Melody Train. Playthings, February 8, 1967, p. 108.

MICKEY MOUSE/DONALD DUCK HANDCAR BASE PROTOTYPE: c.1955. 21 3/4" long.

Except for minor differences, the illustrations on the tin bases of both the lithographed production toy and the handpainted prototype are similar. One notable difference in the prototype, absent from the production toy, shows two Donald Duck nephews dancing in the left foreground. Another difference, also on the left side of the prototype, shows one of Mickey Mouse's nephews sailing a boat. In the production toy, a tunnel has been placed over the spot with this nephew and all that remains is the boat. A third difference is the addition of a semaphore and a crossing gate on the production toy.

Value: ? ?

MICKEY MOUSE KRAZY KAR: 1960s-1970s. 6 1/2" high, 6 1/2" diameter. Battery-operated.

This is a Marx Hong Kong plastic toy in which Mickey sits comfortably on a seat on top of a circular base. With his jointed arms, he moves the Krazy Kar by rotating a wheel on each side of the Kar. The Kar moves in an erratic (not straight) fashion by battery-power.

The smiling Mickey wears his usual short red pants and brown shoes, but also a yellow shirt. The wheels on each side of Mickey are red and white, and the base is yellow. The Kar seat is blue. A paper label on the front of the base reads "Walt Disney's Mickey Mouse Krazy Kar" in multi-colored letters. Underneath the base is another paper label with the Walt Disney copyright and "Walt Disney Productions." Also under the base is an embossed Marx logo.

In addition to the red plastic wheels on either side of Mickey, there are two other small yellow plastic wheels under the base, plus a pair of pivoting black rubber wheels which cause the Kar to move erratically.

The toy was developed by former Marx employee, Cal Cook. (See *GREENBERG'S GUIDE TO MARX TOYS,* Volume I for information on Cal Cook.)

The original box reads "Mickey Mouse Krazy Kar" and Mickey's smiling face is depicted in multi-colored circles.

Value: Excellent with box $95-$110

MICKEY MOUSE LITTLE BIG WHEEL #4671HK: 1973. 10" long. Battery-operated.

In this Marx plastic toy, styled after the famous full-sized, child riding Marx Big Wheel, Mickey rides a tricycle with a low-slung seat, a large front wheel and two small rear wheels.

Mickey, dressed as in the previous toy, has a yellow shirt, red pants and brown shoes. The Little Big Wheel is blue and white. The plastic wheels have silver hubcaps and black tires. The battery-powered Little Big Wheel can go forward or reverse and it also makes a motor noise. The toy was advertised in the 1973 *MARX TOY CATALOG, p. 38* (with no price indicated).

The original box reads "Mickey Mouse Little Big Wheel" and also indicates "Motor Noise." Mickey is illustrated racing along on the Little Big Wheel while Jiminy Cricket runs after him.

Value: Excellent with box $250-$275

Marx 5" plastic Mickey Mouse/Donald Duck Handcar was originally manufactured in 1954 with flanged wheels and came with 2-rail grooved track. In 1955, it was made with non-flanged wheels and came with a channeled metal base. Samuel Caley collection, Robert Grubb photograph.

The plastic Marx Mickey Mouse Krazy Kar was manufactured in Hong Kong circa the 1960s-1970s. It is battery-powered. Samuel Caley collection, Robert Grubb photograph.

MICKEY MOUSE MOTORCYCLE: 1950s. 3 1/2" long. Friction.

This nice little Linemar toy has Mickey riding a tin yellow motorcycle. Mickey is tin, also except for his black rubber ears. He wears a red sweater, long blue pants (instead of his usual short pants) and tan boots. The motorcycle's red and white tin wheels have lithographed spokes. There is also a small black rear rubber wheel on each side.

Donald Duck and Pluto both ride similar motorcycles.

Value: Excellent $225-$250

MICKEY MOUSE MOUSEKETEERS MOVING VAN: 1950s. 13" long. Friction.

This Linemar tin vehicle is lithographed with various Disney characters. The red roof has an illustration of Mickey Mouse. The left side of the greenish-blue and white van reads "Mickey Mouse Mousketeers/Moving..." and shows Mickey, Minnie and Donald. The right side also reads "Mickey Mouse Mouseketeers/Moving..." and shows Pluto, Goofy and Daisy Duck. The van's double rear doors open.

The original box repeats the toy's name "Mickey Mouse Mouseketeers Moving Van." The vehicle is depicted driving down a country road, a house and fence in the background.

Value: Excellent with box $700-$1,000

Mickey Mouse Mouseketeers Moving Van, a 1950s Linemar tin friction vehicle measuring 13" long. Courtesy NEW ENGLAND AUCTION GALLERY, W. Peabody, Massachusetts.

Linemar 1950s tin wind-up toy in which Mickey rocks back and forth on Pluto. Toy is 4 1/2" long. Pluto's rubber tail replaced on example shown. Courtesy LLOYD RALSTON GALLERIES, Fairfield, Connecticut.

MICKEY MOUSE ON PLUTO ROCKER: 1950s. 6 1/2" long. Wind-up.

This is a desirable, attractive and expensive Linemar toy in which Mickey rides a Pluto rocker. Mickey and Pluto are made of tin with rubber-like ears. Pluto has a rubber tail while Mickey is believed to have a string tail.

Mickey rocks back and forth on the Pluto red rocker. Pluto is three-dimensional but his feet are lithographed. Mickey rides on him like a rocking horse and is three-dimensional, also, except for his lithographed legs.

The primarily red box for the toy reads in white letters "Rocking Mickey Mouse on Pluto." The Pluto Rocker is illustrated with a stylized white edge which makes it somewhat difficult to tell that it is a rocker. Both Mickey and Pluto have very enthusiastic expressions.

Value: Excellent with box $1,500-$1,700

This 1950s to 1960s mostly tin Mickey is 6" high. The unicycle moves forward and turns. Courtesy NEW ENGLAND AUCTION GALLERY, W. Peabody, Massachusetts.

MICKEY MOUSE ON SCOOTER: 1950s. 4 1/2" high. Wind-up.

This is a rare and charming Linemar tin toy in which Mickey appears to push the scooter with his right leg while his head moves. The scooter is blue and the wheels are black tin. Mickey's head is larger than usual and his face bears a mischievous expression. Also unusual are his high-waisted pants.

The hard-to-find original box reads "Mickey Mouse On Scooter" and Mickey is illustrated riding the scooter. In addition to the usual markings of the Linemar name, logo and Walt Disney copyright, this box has both Linemar slogans: "Linemar Best By Far" and "One Of The Many Linemar Toys/ Do You Have All Of Them?"

Value: Excellent $700-$800

MICKEY MOUSE RACING KART: 1970s. 5 3/4" long. Friction.

In this plastic Marx toy, Mickey drives a 4-wheeled racing vehicle with black rubber wheels.

The original box reads "Mickey Mouse Racing Kart" and Mickey is illustrated driving the vehicle along a tree-lined street.

Value: Excellent with box $200-$250

MICKEY MOUSE ROLLER SKATER: 1950s. 6 1/2" high. Wind-up.

This is a fine Linemar toy in which Mickey roller skates along by pushing with his right foot. Mickey is tin with rubber ears and cloth pants. The toy comes with a Linemar paper label on the base of one foot.

The colorful original box reads "Mickey Mouse Roller Skater." Mickey is illustrated skating down a flower-lined street watched by Pluto and cheered by Donald.

Value: Excellent with box $1,500-$2,000

MICKEY MOUSE TWIRLING TAIL TOY (See plastic Whirling Tail Mickey Mouse.)

MICKEY MOUSE UNICYCLIST: 1950s-1960s. 6" high. Wind-up.

This is a Linemar toy in which Mickey peddles a unicycyle which moves forward and turns. It is made of tin except for Mickey's rubber ears and cloth pants. The unicycle is mostly red and black with yellow tires and black zig-zag lines representing treads. Goofy rides a similar unicycle.

The original box reads "Mickey Mouse The Unicyclist" and Mickey is illustrated riding the unicycle while Minnie Mouse watches him. Donald Duck is also depicted peddling a unicycle, but such a toy has not yet been seen.

Value: Excellent with box $1,500-$2,000

MICKEY MOUSE WRIST-WATCHES

Marx Japan made children's play watches in various versions including Mickey, Donald and Pluto versions. The watches were plastic, 1 5/8" in diameter, with straps 7 1/2" long made of vinyl with metal clasps. The arms (or feet) of the character point to the time. Different versions of the watch were shown in the *MARX CATALOGS* for 1973, 1974 and 1975 (with no prices indicated).

MICKEY MOUSE WRISTWATCH #6979J: 1973. Wind-up.

The watch was packaged in a blister card marked "Mickey Mouse And His Friends/Toy Wristwatch." However, it is the Donald Duck version that is shown in the 1973 *MARX CATALOG* and described as "Ticking wind-up, open face, see-thru (sic) dial, moving gears." Other wording on the card is "Tick Tock/Wind It Up/Watch It Run/Cannot Be Overwound." The watch is tan colored.

The card is rather exotically colored in purple with green edges. The card's lettering is yellow, orange and black. Illustrated on the card are Mickey, Donald and Pluto. Mickey comments to Donald in a comic-style bubble "Wow Donald, that's some watch!" Donald says "Listen to it tick, Pluto."

Value: $150-$200

MICKEY MOUSE WRISTWATCH #7027J: 1974. Wind-up.

This watch is similar to the 1973 version except that it is red and, in addition, it glows. Only Mickey Mouse is mentioned as being made in this version in the 1974 *MARX CATALOG.*

The watch came on a blister card titled "Mickey Mouse Toy Wristwatch by Marx." Other wording on the card is "Glows In The Dark/Actually Keeps Time/Wind It Up, Watch It Run/Sounds Real/Moving Feet."

The card is dark blue with pale blue and yellow edges and the lettering is mostly white and yellow. Mickey is illustrated standing under a lamp looking at his watch. In a bubble over his head is "Hold under a light. Watch it glow."

Value: $150-$200

MICKEY MOUSE WRIST WATCH #5462J: 1975. Wind-up.

Although very close in appearance to the 1974 version, this watch is advertised as "New" in the 1975 *MARX CATALOG.* The catalog describes it as "Wind-up ticking sound, actually works. Printed face design with moving feet action."

This watch is also red, but unlike the 1974 version, it does not glow. It probably was also sold on a blister card, but the card is not shown in the catalog.

Value: ? ?

MICKEY THE MAGICIAN: Late 1950s. 10 1/2" high. Battery-operated.

In one of Linemar's best and most clever toys, Mickey the Magician makes a chick disappear and then reappear. When the battery-powered action begins, the chick chirps. As described in detail by Craig Ullery in the *ANTIQUE TOY WORLD,* August 1986, "Mickey raises his top hat to reveal the chick. As he lowers the hat. he points the magic wand and then raises his hat to show that the chick has disappeared. He then lowers his hat and raises it again to reveal the chick."

The"magic" is really an electromagnet hidden inside the top of Mickey's hat. The chick is attracted to the magnet and seems to disappear when the hat is raised. When the hat is lowered, the magnet is turned off and the chick reappears.[5]

Mickey is tin with a rubber nose, rubber ears and ceramic hands. His top hat is tin, his coat is woolen and his cloak (cape) is cloth. The chick is tin and sits on the cloth cover of Mickey's performing table, which in turn rests on a metal base.

Mickey is smartly dressed in lithographed striped red and yellow pants, white waistcoat with red and black trim, black bow-tie, black hat and black shoes. His woolen jacket is black with yellow lapels and his black cloth cloak is lined in red.

The chick is, of course, yellow with an orange beak. There is a black cloth cover on the performing table, but in addition, the table has a lithographed red cover with yellow fringes. On the table front is "The Great Mickey" in multi-colored letters. The red, white and green platform of the table is edged with yellow and white stars. The front of the base reads "Mickey The Magician" in black and white letters. The top has a multi-colored diamond pattern and the edges are red with yellow trim. Unfortunately, the chick is frequently missing from this elaborate toy.

Linemar also manufactured a rare mechanical version of Mickey The Magician which has a slightly different appearance. See next listing below. In addition, the manufacturer

Mickey performs tricks on TV for Durham

Mickey performs tricks for interested audience.

NEW YORK — Mickey Magician, one of the three hottest items at the '76 Toy Fair, will be launched nationally this Fall via a major TV campaign.

Aimed at the four to eight age group, Mickey Magician performs 22 different tricks without the aid of motors or batteries.

Made of a durable high-impact plastic, the item stands more than 12-inches high, and comes complete with tricks, magic hat, magic wand, magic cloth, magician cape and storage chest.

Consistent with its staying power strategy, Durham plans to limit the number of Mickey Magicians they make this year, and to promote heavily to achieve an absolute sell-through.

The TV campaign, set to break in September, reflects the same long-term strategy. It will hinge on spot TV in key markets, backed up with national syndication on the Mickey Mouse Club show in addition to network TV support.

The first flight of commercials will break in the key markets in mid-September, and run 12 weeks. A second flight will run 10 weeks beginning late September. The spot campaign will reach 30 top toy markets.

The manufacturer Durham's version of Mickey The Magician debuts on TV. Playthings, June 1976, p. 46.

Durham made a plastic Mickey Magician in the mid-1970s. Somehow this Mickey, despite being non-powered, manages to perform several tricks.

The original box for the Linemar battery-powered Mickey reads "Mickey The Magician" with 'Magician' in uneven lettering. The toy is depicted realistically on the box, with the chick on the table making a "chirp chirp" sound. The only major difference between the box illustration and the actual toy is that in the illustration, the base with the words "Mickey The Magician" is lacking.

The sides of the box show three illustrations of Mickey making the chick disappear and reappear. The top of the box has the number "J-4126" along with a depiction of Mickey with a wand in his left hand and his hat in his right hand.

Value: Excellent with box $2,500-$3,200

Linemar 1950s battery-operated Mickey The Magician is 10 1/2" high. Mickey makes a chick disappear and reappear. David Allen collection, Robert Grubb photograph.

MICKEY THE MAGICIAN: 1950s-1960s. 6 1/2" high. Wind-up.

This rare Linemar wind-up toy was described in *HAKE'S AMERICANA & COLLECTIBLES CATALOG* for November 1986 as follows: "Mickey holds a hat over the box in front of him. He raises and lowers his arm and beneath the hat alternatingly appears either a small colorful bird or a white rabbit."

The toy is tin but Mickey's nose and ears are rubber. Mickey wears red and yellow pants, a black and white tuxedo jacket and a pink bow-tie. On the front of the box/table is "Mickey The Magician."

This toy is not quite as elaborate as the battery-operated version described immediately above. Instead of the woolen jacket and cloth cloak of the battery-operated version, this toy has lithographed clothes and no cloak. He uses a much smaller hat than the battery-operated version's top hat and there is no pedestal under the table, so that it sits directly on the base. However, in the battery-operated version, Mickey only produces a chick while in this wind-up version, Mickey performs magic with a bird and a rabbit. The wind-up version is considerably rarer than the battery-operated version.

The original box for the toy reads "Mickey The Magician" with the same uneven lettering used for 'Magician' as on the battery-operated version's box. Mickey's illustration shows him with the wand in his left hand and the top hat in his right hand. As already pointed out, the actual toy does not have a top hat (or cloak). The rabbit is shown sitting on the table.

Value: $4000-$5000

MICKEY THE MUSICIAN: 1950s. 12" high. Wind-up.

In this Marx tin toy, Mickey's head moves as he plays the xylophone. Mickey is plastic and the rest of the toy is tin. The xylophone has been found in brown or bluish-grey and may have come in additional colors. The instrument has a strange appearance for there is a gap in the middle of the keys. It looks as if a note is missing, but it was manufactured with just two keys on each side and a space in between. The xylophone stands on a mostly red lithographed carpet with an ornate design resembling that of a Persian carpet. The rest of the base is yellow with a red rim.

Linemar also manufactured an all-metal, more colorful, smaller version of this toy (see following listing). However, the Marx version is harder to find than the Linemar version.

The original box for the Marx toy reads "Mickey The Musician/I Play the Xylophone" in red and black letters. Mickey is illustrated enthusiastically playing the instrument while wavy lines and notes suggest music emanating from the instrument to surround Mickey.

Value: Excellent with box $600-$700

MICKEY WITH XYLOPHONE: 1950s-1960s. 6 1/2" high. Wind-up.

This is the Linemar version of the toy described above. Again, Mickey plays the xylophone. The toy is tin except for Mickey's rubber tail. Mickey's xylophone has five keys colored respectively green, pink, orange, blue and yellow. The xylophone is mostly red, white and yellow, basically matching the lithographed carpet beneath it. Mickey and the carpet are on a yellow base.

This Linemar toy is 6 1/2" high, about half the size of the Marx 12" high Mickey The Musician toy. Mickey is metal here instead of plastic and has a more colorful xylophone. The carpet is similar to the Marx version. In addition, the Linemar version has five keys instead of four, eliminating the problem of the Marx version, where a key appears to be missing in the center of the xylophone. The Linemar version is easier to find than the Marx version.

The original box for the toy reads "Mickey Mouse With Xylophone" and shows Mickey playing a xylophone with striped keys, while other Disney characters such as Goofy, the Three Little Pigs, Minnie Mouse and Donald Duck's nephews play various instruments in an orchestra.

Value: $600-$700
With box $1,500-$1,800

The 6 1/2" high tin Linemar wind-up Mickey Mouse With Xylophone was manufactured circa the 1950s-1960s. Courtesy NEW ENGLAND AUCTION GALLERY, W. Peabody, Massachusetts.

WALKING MICKEY MOUSE: Early 1970s. 3 1/4" high. Wind-up.

In this inexpensive Marx plastic walker toy, Mickey wears a big smile and his hands rest on his stomach. Although a wind-up, this toy's shape, with its movable legs, resembles those of the ramp walkers (see later in this chapter).

Value: ? ?

WHIRLING TAIL MICKEY MOUSE: 1950s. 7" high. Wind-up.

In this Marx toy, Mickey stands with legs apart and hands on hips. He turns around while his tail whirls. Mickey is plastic with a metal tail (See next toy listing for Linemar tin version of this toy.).

Mickey wears his usual red pants, but his gloves and shoes are yellow. Mickey also comes in a rarer yellow pants version.

The original box reads "Whirling Tail Mickey Mouse." It is plain, without illustration. However, the same toy comes, much less commonly, in an illustrated box marked "Mickey Mouse Twirling Tail Toy By Marx." Mickey is depicted walking along, taking a rather large step, while his tail whirls. The thumb of his raised left hand points to the name "Marx." (The Linemar tin version came in yet another box.)

Value: Excellent with plain unillustrated box $250-$350

WHIRLING TAIL MICKEY MOUSE (More common version): 1950s. 6" high. Wind-up.

The Linemar tin Whirling Tail Mickey Mouse is similar in appearance and action to the Marx plastic version described above, but he is one inch shorter, his arms are closer to his body and his yellow shoes are molded together. Mickey is tin with rubber ears and a vinyl tail.

The original yellow box is marked "Whirling Tail Mickey Mouse" and Mickey is depicted with his left hand on his hip and his right hand pointing to the box's title.

Value: Excellent with box $900-$1,000

WHIRLING TAIL MICKEY MOUSE (Scarce Version): 1950s. 5 1/2" high. Wind-up.

The scarce version of the Linemar tin Whirling Tail Mickey Mouse is similar to the more common version of the toy described directly above. However, the *NEW ENGLAND AUCTION GALLERY CATALOG,* July 1992, West Peabody, MA, describes important differences in the rare version, such as Mickey having a different face, differently colored feet (black), being thinner but more pot-bellied and being a half-inch shorter.

The original box has the same wording as the more common version, namely, "Whirling Tail Mickey Mouse," but differs in that the box is blue instead of yellow and Mickey is shown with his hands outstretched on either side of him.

Value: Near mint with box $1600-$2000

NOTE: For additional Mickey Mouse toys, see previous listings under "Disney Toys."

MINNIE MOUSE: 1958. 6 1/2" high. Wind-up.

This delightful Linemar toy has Minnie Mouse rocking in her chair while knitting. Each knitting needle actually moves, one after the other. As she rocks, Minnie's head moves back and forth and her feet vibrate. The toy is tin except for Minnie's rubber ears and the woven material that she "knits" as she rocks.

Minnie has a big smile on her flesh-colored face and her eyes boast long eye-lashes. On her head is a lithographed yellow bow. She wears a red and yellow plaid blouse with white collar and cuffs and a red skirt with black stripes and white scalloped edges. Minnie's arms are lithographed, but her legs are three-dimensional. Her pointed red shoes have little pink bows. Minnie's ornate rocking chair is yellow, blue, white and black, and has a scalloped design on the sides. The front and back of the chair have a flowery design. On the back is the Linemar name and logo and the Walt Disney Productions copyright. This version is the best known Linemar example of a Minnie Mouse toy. One problem with the toy is the difficulty in winding it due to the winding key being rather close to the mechanism which causes the chair to rock.

The toy is easily available and one of the more reasonably priced Linemar wind-up toys, however it can be difficult to find the box.

The most available version of the Whirling Tail Mickey Mouse toy was made by Linemar in the 1950s. The 6" high toy is tin with rubber ears and a vinyl tail. Courtesy LLOYD RALSTON GALLERIES, Fairfield, Connecticut.

The original box reads "Minnie Mouse" and in smaller letters "Watch Minnie Rock In Her Rocking Chair And Knit." Minnie is illustrated knitting in her rocking chair, an accurate representation of the actual toy, except for the different pattern of Minnie's clothes.

The Minnie Mouse toy was described in the 1958 *SEARS CHRISTMAS CATALOG,* p.366, as "Minnie rocks and knits. Head and legs move. Brightly painted metal. Enclosed bottom." It was priced at 89 cents!

Value: Excellent $400-$600
Excellent with box $900-$1,100

Linemar 1958 tin wind-up. Minnie Mouse knits as her chair rocks. Samuel Caley collection, Robert Grubb photograph.

NODDERS: 1960s. About 2 1/4" high each. Non-powered.

Marx Hong Kong made this plastic toy in which the character's head nods. The toy was made in versions with Bambi, Donald Duck, Dumbo, Goofy, the Mad Hatter, the March Hare and Pluto. Except for Bambi and Pluto, who have all four legs on the ground, the other figures are formed in bent-over positions which may help balance them.

Each toy is embossed on the sides with "Walt Disney Productions," the copyright symbol and "Hong Kong." The Marx name is not seen, but it is thought that these toys originally came with a paper label on the base that may have had the Marx name.

BAMBI:

The toy is primarily yellow with brown, white and red accents. Bambi's stomach and the inside of his ears are a bright pink which clashes somewhat with his yellow body. An appealing touch is the deer's long eye lashes. The toy is made of a lighter weight plastic than that of the other toys in the series but its identical markings show that it is related to the others.

DONALD DUCK:

The white Donald is dressed in a blue sailor suit with red and white trim and a blue cap.

DUMBO THE ELEPHANT:

The dark grey Dumbo wears a red, white and blue collar and a red and yellow hat. He is the sturdiest of the varied toys in the series.

GOOFY:

Goofy has a flesh-colored face, big black nose and black ears. He wears a red and black hat, red shirt, black waistcoat, blue pants and has large brown shoes.

THE MAD HATTER:

The Mad Hatter is a character from the 1951 Disney film "Alice In Wonderland." He is buck-toothed, has blonde hair and wears a large blue top-hat. His coat is yellow, his stand-up collar is white and his pants and oversized shoes are blue. He holds a cup and saucer in his hands.

THE MARCH HARE:

The March Hare, another character from the 1951 "Alice In Wonderland" film (along with the Mad Hatter above) is brown, buck-toothed, with a red nose, yellow thatch of hair and a silly expression. He wears a red coat, large yellow bow-tie, blue pants and has oversized blue shoes.

MICKEY MOUSE:

Mickey Mouse, the most often seen toy in the series, has a flesh-colored face and black ears. He wears a blue, red and black outfit and - differently from the other toys - he holds a large black top hat which rests on the ground.

PLUTO:

The yellow Pluto has black ears, a black tail and black, white and red trim.

Value: $15-$20 each.

Measuring 2 1/4" high, the plastic Mickey Mouse is the only Nodder in the Marx Hong Kong series to hold a hat. Ed Owens collection.

The 1960s Marx Hong Kong Nodders (Bambi, Dumbo, Mad Hatter and the March Hare) are plastic and measure about 2 1/4" high each. Philip Rolin collection, Robert Grubb photograph.

ONE HUNDRED AND ONE DALMATIANS

Linemar manufactured two toys based on the 1961 Disney animated film "One Hundred And One Dalmatians." Each is made of plush on a frame body.

LUCKY PUP: 1961. 5 1/2" long. Wind-up.

According to *PLAYTHINGS MAGAZINE* for April 1961, p. 117, Lucky Pup "wags his tail and raises to a begging position, with front paws and head moving."[6] Toy is black and white, naturally, and he barks, too.

The original selling price was 79 cents.

Value: Excellent with box $175-$200

WIND-UP DISNEY DALMATIAN INTRODUCED BY LOUIS MARX

Lucky Pup sits up and begs, barks, and wags his tail—with just a turn of the key. Made by Louis Marx, the toy is a mechanical wind-up replica of the star of

Walt Disney's new cartoon feature, "101 Dalmatians." Composed of realistic plush on a sturdy frame body, Lucky Pup measures 5½"x3"x3½" high.

Powered by a strong clock spring motor, he wags his tail and raises to a begging position, with front paws and head moving. Individually boxed, the toy retails for 79c and is packed three dozen to a carton.

A character from the Disney animated film "One Hundred and One Dalmations," the wind-up "Lucky Pup" is 5 1/2" long. *Playthings Magazine*, April, 1961, p. 117.

The 1955 Marx Pecos Bill Ridin' Widowmaker is a 9 3/4" high plastic wind-up with a tin lasso. As the horse turns in a circle, the lasso vibrates. Courtesy SMITH HOUSE TOYS, Eliot, Maine.

SLEEPY LUCKY PUP: 1961. 7" long. Wind-up.

Sleepy Lucky Pup is not as energetic as his brother, Lucky Pup. He merely lays stretched out, sleepily moving his head and legs. Pup is black and white, of course.

The original box reads "Sleepy Lucky Pup From Walt Disney's 101 Dalmatians." The Letter 'U' of "Lucky" is shaped like a horseshoe. The pup is illustrated sleeping contentedly.

Value: Excellent with box $175-$200

NOTE: Also see 101 Dalmatians Snowdomes in Snowdomes section.

PECOS BILL RIDIN' WIDOWMAKER: 1955. 9 3/4" high. Wind-up.

Cowboy Pecos Bill and his horse Widowmaker are characters from the 1948 Disney short film "Melody Time." The toy Pecos Bill, twirling a lasso, sits on his rearing horse Widowmaker, while the horse turns around in a circle. The toy is plastic except for Pecos Bill's metal lasso.

Pecos Bill has brown hair, a brown neckerchief and brown gloves. He wears a red bolero and his pink shirt is the same pink as is used for his skin. A black cowboy hat, belt, holsters and boots complete his costume. His horse Widowmaker is brown with a yellow mane.

The Marx logo is embossed on the cowboy's left leg. The Disney copyright symbol and "W.D.P." are embossed on the horse's left flank.

The plastic Pecos Bill Ridin' Widowmaker is the same basic shape and has the same action as the earlier 1938 tin Lone Ranger toy, but the Lone Ranger toy is almost two inches smaller and that cowboy has a jointed left arm.

The original box reads "Pecos Bill Ridin' Widowmaker" and has an accurate drawing of the actual toy. The box sides depict a bow-legged Pecos Bill with his lasso around his name while the horse watches. Under his Name is "He's Ridin! He's Ropin!" (sic). The box top has instructions on how to insert the lasso into the toy. The 1955 *SEARS CHRISTMAS CATALOG* advertised the toy for 94 cents.

Value: Excellent with box $250-$275.

PINOCCHIO #J-9151: 1962. 5 3/4" high. Wind-up.

Pinocchio is the central character in the wonderful 1940 Disney film of the same name. In this charming Linemar tin toy, Pinocchio moves his arms up and down and walks by carefully stepping one foot over the other.

Pinocchio has a smiling pink-cheeked face, large blue eyes and black hair. He wears a pale yellow shirt with a white collar, big blue bow, black bolero with pale yellow trim and red pants. His hat is dark yellow with a lithographed blue band within which is stuck a red feather. Pinocchio's gloves are white, but he wears no shoes. His brown legs and feet are shaped like those of many robots, rectangular with finger-like projections.

On Pinocchio's back are the Linemar logo and both the Linemar and Walt Disney copyright symbols, along with the stop and go lever.

During 1939-1940, Marx manufactured a number of Pinocchio toys, the closest to the Linemar version being the Pinocchio Walker. However, the Marx Pinocchio Walker, at 8 1/2" high, is taller, has moving eyes and lacks the bolero worn by the Linemar version. (For information on Marx Pinocchio toys, see *GREENBERG'S GUIDE TO MARX TOYS,* Volume I, pps.82-85.)

Linemar Pinocchio 1962 tin wind-up moves his arms and walks. Dr. Malcolm Kates collection, Robert Grubb photograph.

The very attractive original box for the Linemar toy reads "Walking Walt Disney's Pinocchio." He is illustrated with a book satchel on his back and an apple in his right hand, walking on a cobble-stone street. Jiminy Cricket walks beside him while the moon shines down on them.

Pinocchio was introduced briefly in the April 1962 issue of *PLAYTHINGS.* J-9151 is the catalog number given for the toy.[7]

Value: Excellent with box $750-$1,000

PINOCCHIO ON WHALE: Mid-1960s. 4 1/2" long. Friction.

This is an unusual Marx plastic toy in which Pinocchio rides a smiling whale. There is a paper Marx sticker underneath the whale.

As described in *HAKE'S AMERICANA & COLLECTIBLES CATALOG,* York, Pa, September 1990, the original box does not use Pinocchio's name but reads "Walt Disney's Friction Toy" and "Fast Hill Climbing/Precision Vehicle/Painted By Hand By Artists." The Magic Marxie figure is shown on the end flap. The box is a display type with a cellophane window. Inside is an illustrated cardboard insert showing the water and shore.

PLUTO TOYS

Pluto originally started out as a bloodhound in the Disney short "The Chain Gang." By the time the Disney cartoon "Playful Pluto" was released in 1933-1934, Pluto had developed into the dog we know and love, and many excellent toys were manufactured in his likeness. Pluto is usually yellow or yellow-tan with a red collar.

MUSICAL PLUTO HONEYMOON EXPRESS: 1960s. 8" x 8" base. Wind-up.

This is a Marx Japan toy in which Pluto speeds around the circular track of the Honeymoon Express-type toy, while musical notes sound. The toy is tin except for Pluto's rubber tail.

The very attractive and uncommon toy is colored mostly in pastel shades of pink, green, blue and white. Flowers are everywhere and almost every major Disney character is lithographed on the toy. In one corner of the base, Mickey plays a drum; in another corner, Goofy plays an accordion; in a third corner, Donald holds a violin; while in the fourth corner, Pluto plays a xylophone. In the center of the base are characters not commonly seen, such as the White Rabbit and Jiminy Cricket, along with a Chipmunk and one of Donald Duck's nephews. On the base are three pink tunnels through which Pluto passes, one with a wood pattern, one lithographed with birds and flowers, while the last and biggest is illustrated with Snow White and the Seven Dwarfs. This tunnel has the wind-up key and the name "Musical Pluto" lettered on the top. On the sides of the base "Musical Pluto" is repeated while the remaining two sides show Mickey walking with Pluto. Under the base are the Marx name and logo, the Walt Disney Productions copyright and "Made In Japan."

The original box is rather unexciting. It reads "Pre-School Toy/Disney Musical Choo Choo" and the toy is shown.

Value: Excellent with box $500-$750

This 1960s tin wind-up Marx Japan Honeymoon Express-type toy measures 8" x 8". Pluto speeds around a circular track while musical notes sound. Samuel Caley collection, Robert Grubb photograph.

PARTY PLUTO: 1954. 10" high. Wind-up.

According to *PLAYTHINGS* magazine, April 1954, the Marx Party Pluto "...is a master of motion as he dances, rocks back and forth, shakes bells and blows a horn."[8] The toy is plastic except for a metal bell held in each front paw and his rubber ears and tail. It had a suggested retail price of $2.98 in 1954.

Pluto's small top hat is green with a red stripe, matching the green, red-tipped horn that he blows. Linemar also made a tin version of this toy in which Pluto holds a baton in one paw instead of bells in both paws. The Marx plastic version is rarer than the Linemar tin version (See "Pluto Drum Major" later in this section.)

Value: $275-$400

DISNEY'S PARTY PLUTO BY LOUIS MARX CO.

Walt Disney's Party Pluto, manufactured by Louis Marx and Company, 200 Fifth Avenue, New York City, is a master of motion as he dances, rocks back and

forth, shakes bells, and blows a horn.

Pluto's rubber ears and tail also complete the moving picture. A bright green top hat, vivid red collar, and green horn complement Pluto's yellow body.

A wind-up key sets him in motion and his actions are controlled by a lever. Suggested retail price is $2.98.

The 1954 mostly plastic Pluto is a 10" high wind-up. He shakes bells in both paws. Description and photo from Playthings, April 1954, p.126.

PLAYFUL PLUTO: 1950s-1960s. 6" long. Wind-up.

This is a Linemar toy in which Pluto vibrates as his head nods and his tail whirls. The toy is tin except for Pluto's rubber tail. Playful Pluto was sold in a separate box as well as packaged together with Goofy (see listing following this one).

The original box reads "Playful Pluto" which is the name of the early-1930s Disney cartoon which brought Pluto to prominence. Pluto is depicted walking, his head turned back to watch his whirling tail. A smiling Mickey Mouse and admiring Daisy Duck watch the action.

Value: Excellent with box $450-$650

The 6" long Linemar tin wind-up Playful Pluto vibrates as its head nods and tail whirls. Courtesy NEW ENGLAND AUCTION GALLERY, W. Peabody, Massachusetts.

PLAYFUL PLUTO AND GOOFY: 1950s-1960s. Pluto: 6" long, Goofy: about 5" high. Wind-up.

The Linemar Playful Pluto (described above) and Goofy were sold together in the same box as well as being packaged separately. The characters are both tin and both have rubber tails.

The toys are not hard to find, but the box is rare. It reads "Playful Pluto And Goofy" with the names being shown on a sign post. Both Pluto and Goofy are depicted in similar style to the illustrations on their respective boxes when they are packaged separately.

For photo and information on the Goofy toy, see the first listing in the Goofy section.

Value: Excellent with box $1,400-$1,500

PLUTO CAR: 1950s. 6" long. Friction.

This is a Linemar toy in which a celluloid Pluto drives a tin Jaguar-type car.

Value: ? ?

PLUTO CROUCHING: 1960s-1970s. 6" long. Friction.

This is a Marx plastic toy in which Pluto is molded with his chin on the ground and his rump in the air. See next listing for larger pull-toy version.

Value: $225-$400

PLUTO CROUCHING: 1960s. 18 1/2" long. Non-powered pull-toy.

This is another Marx plastic toy in which Pluto is molded with his chin on the ground and his rump in the air. However, it is a pull-toy and over three times longer than the 6" friction version described immediately above. There is a decal on the toy's back which reads "Pluto/Squeeze Him/Pinch Him."

The toy was advertised in the 1967 *MARX CATALOG OF TELEVISION PROMOTIONAL ITEMS* (p.39) and indicated that "When you squeeze Pluto, he barks through his platform sound mechanism. He rolls along on 4 rigid [black] wheels and is just right for toddlers to pull along by his collar string loop." No price was given. The toy was still being advertised seven years later in 1973.

Value: $400-$500

PLUTO DRUM MAJOR: 1950s. About 7" high. Wind-up.

In this Linemar toy, Pluto rocks back and forth as he shakes a bell in his left front paw, moves a baton in his right and blows a horn. The toy is tin except for Pluto's rubber tail.

Pluto's small green top hat has a red band and matches the green, red-tipped horn he blows. This Linemar Pluto is similar to the Marx Party Pluto (previously described) but at 7" high, it is 3" shorter than the Marx toy, and it is made of tin instead of plastic. The Pluto Drum Major also has a baton in one paw instead of bells in both front paws. The Linemar Pluto Drum Major is easier to find than the Marx Party Pluto.

Linemar 7" high tin wind-up Pluto Drum Major rocks back and forth as he plays his instruments. David Allen collection, Robert Grubb photograph.

The very attractive original box reads "Pluto The Drum Major" and around the sides of the box are different views of Pluto playing his instruments. Adding to the box's appeal is its striped floor and striped top.

Value: $300-$350
Excellent with box $500-$800

PLUTO LANTERN: 1950s-1960s. 7" high. Battery-powered.

This battery-powered Linemar Pluto Lantern has a transparent glass center in which a bulb lights. Pluto's head, shoulders and torso, shown down to below the waist, are tin. His ears may be rubber.

A similar lantern was made with Popeye, but the Pluto version is less common. The original box reads " Pluto Lantern" and the lantern is depicted. Strangely, Mickey's head looms under the title, longer than Pluto's tail.

Value: ? ?

A 1950s-1960s 7" high battery powered Pluto Lantern. The lantern's center lights up. Courtesy SMITH HOUSE, Eliot, Maine.

PLUTO MOTORCYCLE: 1950s. 3 1/2" long. Friction.

In this Linemar tin toy, Pluto, dressed in yellow shirt, long blue pants and yellow boots, rides a mostly tan motorcycle. The red and white tin wheels have lithographed spokes. There is a small rear black rubber wheel on each side. Donald Duck and Mickey Mouse both ride similar motorcycles.

Value: Excellent with box $225-$250

In this 1950s tin friction toy, Pluto and Donald Duck each ride 3 1/2" long motorcycles. A version was also made with Mickey Mouse. Donald Duck Motorcycle by Linemar, shown on the left, Pluto Motorcycle on the right. Courtesy NEW ENGLAND AUCTION GALLERY, W. Peabody, Massachusetts.

PLUTO OLD FASHIONED CAR: 1960s. 3 1/2" long. Friction.

In this plastic Marx toy, a three-dimensional Pluto drives a red car. As described in *HAKE'S AMERICANA & COLLECTIBLES CATALOG,* York, PA, April 1993, the maroon-roofed car has blue, gold and silver trim and is designed to resemble a 1910 Studebaker.

Wording on the box front is "Walt Disney's Driver" along with "Walt Disney Productions" plus the Disney copyright and the Marx logo. The fact that the box title does not specify a particular Disney character indicates that the same box could have been used, and possibly was, for additional characters besides Pluto. The box has a display window and an insert depicting a wood fence with a green background.

Value: Excellent $100-$200

PLUTO PULLING WAGON: 1950s. 9" long. Friction.

In this Linemar tin toy, Pluto pulls a wagon. The outer part of the wagon is red, the inner part yellow. The black tin wheels have blue hubcaps. The Pluto figure in this toy is the same as that used in the Playful Pluto toy, described previously.

Value: Excellent $350-$400

PLUTO SITTING (Small): 1950s. 3" long. Friction.

In this Linemar tin toy, Pluto is in a sitting position with his two front paws stretched out in front of him. His back legs are lithographed but his front legs are three-dimensional. On his left leg is the Disney copyright symbol and "W.D.P." and on his right leg is the Linemar logo. (For photo, see "Thumper Rabbit.")

Value: Excellent $100-$110

PLUTO SITTING (Large): 1950s. 7 1/4" long. Wind-up.

This Linemar tin toy is similar to the one just described, but it is much longer and it is a wind-up.

Value: Excellent $300

(Left) 1950s Linemar Pluto measures 7 1/4" long, is wind-up powered and is all tin. (Right) 1960s Marx Japan Disney Kart measures 6" long and is friction-powered. The Kart is made of tin and Pluto's head is made of vinyl. Courtesy LLOYD RALSTON GALLERIES, Fairfield, Connecticut.

PLUTO TWIRLING TAIL TOY: 1950s-1960s. 5 1/2" high. Wind-up.

In this Marx toy, a plastic standing Pluto moves in a circular pattern as his metal tail twirls.

This toy has been seen in different boxes. The less common box is a plain, unillustrated one lettered "Whirling Tail Pluto." The other box is illustrated with Pluto shown watching his twirling tail and is lettered "Pluto Twirling Tail Toy/By Marx." It is conjectured that the plain box came first and was later changed to an illustrated one in order to stimulate sales. Donald Duck was also packaged in two different boxes with similar titles.

Value: Excellent with box $175-$210

PLUTO WITH WHIRLING TAIL: 1950s-1960s. 4 1/2" long. Wind-up.

This is another Linemar tin toy in which Pluto's tail whirls. This time he is in a standing position and vibrates instead of moving in circles as in the preceding toy. It is believed that Pluto has rubber ears and a rubber tail. This version is not as common as some of the other Pluto toys.

The original box reads "Pluto w/ Whirling Tail" (sic) and Pluto is depicted accurately.

Value: Excellent with box $300-$400

Four Marx plastic wind-up toys manufactured circa the 1950s-1960s, including the Pluto Twirling Tail toy. Heights range from 5 1/2" to 8". Ralph Kaufman collection, Dick Kaufman photograph.

PLUTO UNICYCLIST: 1950s-1960s. 5 1/2" high. Wind-up.

Made by Linemar, the 5 1/2" high tin Pluto is yellow with a red collar. The unicycle is red, yellow and black with lithographed spokes and treads. It rolls forward and turns. Goofy and Mickey Mouse ride similar unicycles.

The multi-color box illustration shows Pluto peddling along rather energetically, while on the right, Donald waves him along. The box title is "Pluto The Unicyclist." Other markings are the Linemar logo and slogan Linemar/Best By Far.

Value: With box $1,500-$2,000

The Linemar tin wind-up Pluto Unicyclist was manufactured in the 1950s-1960s and is 5 1/2" high. Courtesy SMITH HOUSE, Eliot, Maine.

ROLLOVER PLUTO: 1950s-1960s. 6 1/2" long. Wind-up.

This is an unusual Linemar toy in which Pluto runs forward, rolls over and raises up on his hind legs to beg. Pluto is plush covered.

The description and photograph of the toy can be seen in the *CHRISTIE'S COLLECTIBLES AUCTION CATALOG* for March 1989. The photo does not show the box title, but Pluto is illustrated as sitting up and begging.

In 1939, Marx manufactured an 8" long tin Rollover Pluto, but that toy does not raise up on his hind legs to beg.

Value: $300-$400

RUNNING PLUTO: 1950s. 4" long. Friction.

In this Linemar tin toy, Pluto's tongue moves in and out as he moves forward. Despite the name of the toy, Pluto is actually formed in a crouching position.

The original box reads "Running Pluto" and he is illustrated running after a bone. Near his tongue is lettered "Moving Tongue" with the movement being indicated by dotted lines.

Value: With box $300-$350

Despite the name, the tin Linemar Running Pluto actually crouches. The toy was manufactured in the 1950s, measures 4" long, and is friction powered. Courtesy NEW ENGLAND AUCTION GALLERY, W. Peabody, Maine.

WALKING PLUTO (With Moving Head): 1950s-1960s. 5 1/2" long. Wind-up.

In this Linemar toy, Pluto moves his head as he walks along. He has a plush covered body with rubber ears and a rubber tail. On Pluto's stomach is a paper label with the Linemar logo and the Walt Disney Productions copyright.

The original box is lettered "Walking Pluto" and in smaller letters "With Moving Head." A jaunty Pluto is illustrated walking on the sidewalk past a stone wall, with a house and trees in the distance, while a silhouetted young boy looks on. (See also "Walking Pluto With Moving Head and Tail" below.)

Value: Excellent with box $350-$400

The 5 1/2" long plush covered Linemar Pluto was manufactured in the 1950s-1960s. Pluto moves his head as he walks. Courtesy NEW ENGLAND AUCTION GALLERY, W. Peabody, Maine.

WALKING PLUTO (With Moving Head And Tail): 1950s-1960s. 8" long. Wind-up.

This Linemar Pluto moves his head while his tail twirls. He has a plush covered body with a rubber tail. This Walking Pluto differs from the one in the immediately preceding listing in that he has the added tail movement, has a tongue sticking out, and at 8" long is 2 1/2" longer than the preceding 5 1/2" long Pluto. Both toys have moving heads and plush covered bodies in common.

The original box reads "Walking Pluto/Moving Head/ Whirling Tail." Pluto, as illustrated, seems to have a rather long body, like that of a Dachshund, although in the actual toy, the body does not seem any longer than usual.

Value: Excellent with box $400-$450

The 1950s-1960s plush-covered wind-up measures 8" long and has a moving head and tail. Courtesy SMITH HOUSE, Eliot, Maine.

WALKING PLUTO: (actually sitting) 1950s-1960s. 7" long. Wind-up.

Walking Pluto is a misleading name for this Pluto who is actually stretched out in a sitting position. The toy moves by means of wheels under the base. The toy has a nodding head and a tin tongue. His tail is rubber.

Value: Excellent $215-$225

Pluto, rolls while lying down. Linemar 7" long tin wind-up Pluto was manufactured in the 1950s-1960s. The toy's head moves. Courtesy NEW ENGLAND AUCTION GALLERY, W. Peabody, Maine.

WISE ACTION PLUTO: 1950s. 5 1/2" long. Wind-up.

In this uncommon Linemar tin toy, Pluto moves along a table top supposedly without falling off the edge. Like all toys of this type, it has to be watched because, once in a while, it does go off the table edge.

In 1939, Marx manufactured a tin wind-up Wise Pluto but it is larger, being 8" long.

NOTE: For additional Pluto toys, see previous listings under "Disney Toys."

Value: $150-$250

PROFESSOR LUDWIG VON DRAKE: 1960s. 6" high. Wind-up.

Professor Ludwig Von Drake is a humorous duck with glasses and a German accent who, starting in the early-1960s, narrated many of the Disney television programs. As portrayed in this Linemar tin toy, the professor's arms move up and down as he walks forward.

The Professor wears a black top hat, black tie and red and white diamond-patterned vest. His face, hands and shirt are white, his beak and feet are yellow. On the tip of his nose, perches a pair of metal glasses. At each side of his head is a bit of fuzz representing wisps of hair. In the Professor's right hand is a metal pointer, in case he needs to use it in a lesson or lecture. The toy's feet, with their rectangular shape and finger-like projections, are similar to those of the Linemar Pinocchio toy.

The original box for the toy reads "Walking Walt Disney's Professor Von Drake." The old Professor is rendered quite accurately except for his towering black top hat which is rather small on the actual toy.

Value: Excellent with box $650-$700

RAMP WAKERS: 1960s. About 3" high each. Non-powered.

Ramp Walkers, sometimes known as Funny Walking Toys, are small inexpensive Marx Hong Kong toys with movable legs. Although they walk down a ramp or incline, the toys can also walk horizontally across a table if a weight attached to them by a string or thread is allowed to dangle over the table edge. Some of the toys can even walk backwards. The toys were packaged in cellophane with the label "No Winding/Place On Incline And Watch 'Em Walk." The toys are plastic but a few of them may have metal legs.

A 5 1/2" x 8 1/2" advertising and instruction sheet was packaged with some of the toys and states: "To operate, bring thread [or string] back between legs. Face toy away from table edge. Hang weighted end of thread just over edge of table. Gently rock toy and watch it go." As well as giving instructions, the sheet depicts, in color, many different walking toys, including fourteen Disney toys, which are shown at the top of the sheet. The Magic Marxie figure is shown on the bottom of the sheet. Many of the toys described in the listings are shown.

Value: $35-$75

This 6" Linemar 1960s tin wind-up Professor Von Drake moves his arms as he walks forward. Courtesy LLOYD RALSTON GALLERIES, Fairfield, Connecticut.

Walking Toys original artwork by Marx artist Roy Ahlgren.

A 5 1/2" x 8 1/2" color sheet packaged with some Ramp Walker toys. Although the toys are called Funny Walking Toys, Marx used other names, too.[14] Disney toys are shown at the top of the page.

The Ramp Walkers came in the versions listed below.

DONALD DUCK PUSHING A WHEELBARROW

Donald pushes a brown wheelbarrow (also see next listing).

DONALD DUCK - WALT DISNEY WALKING TOY WITH MINIATURE PLUTO

This 1950s toy was made by the Charmore Company, a company owned by Marx. As described in the September 1987 *HAKE'S AMERICANA & COLLECTIBLES CATALOG,* York, PA, the toy consists of the very same Marx Donald Duck walker pushing the wheelbarrow listed above, including a brick to put in the wheelbarrow, plus a small Pluto with metal legs. In addition, the toy comes with its own tin ramp, about 24" long, which depicts an aerial view of an orange road going through farmlands, the suburbs and the city. The ramp shows a lithographed airplane spelling out in sky-writing "Long Walk Ramp for Walking Toys!"

The Charmore Company made a similar type walking toy with a tin ramp and the lengthy title "Watch The Nursemaid Take Dick Tracy's Bonnie Braids For A Stroll." The Marx Company manufactured the same toy with a different title "Nora The Nursemaid/Watch Her Stroll Thru (sic) The Park."

The original box for the Donald Duck - Walt Disney Walking Toy With Miniature Pluto is shown in the September 1987 HAKE'S catalog. The wording is "One Of The Walt Disney Walking Toys/Complete With Ramp." Illustrated on the box are Mickey along with Minnie who pushes a carriage with one of Mickey's nephews in it. Also shown are Pluto and Donald with his wheelbarrow. The box flaps show just the heads of all the characters.

Value: Excellent with box $275-$300

Other Marx Hong Kong Ramp Walkers are as follows:

DONALD DUCK WITH GOOFY

Donald leads the way with Goofy attached to him.

DONALD DUCK WITH NEPHEWS IN WAGON

Donald pulls a red wagon in which his three nephews, Huey, Dewey and Louie, ride.

GOOFY RIDING HIPPO

As the title indicates, Goofy, wearing a red shirt, red hat, blue pants, white gloves and black shoes, sits astride a red and gold blanket on top of a hippopotamus. The toy is embossed on the left side with the Marx logo and the Walt Disney Productions copyright.

JIMINY CRICKET PUSHING BASS FIDDLE

Jiminy holds one end of the instrument while the other end has a wheel on it. Jiminy wears a black coat, blue pants and a blue hat.

Marx Hong Kong plastic Ramp Walker is 3" high and was manufactured in the 1960s. Toy walks down an incline or across a table if weight is attached over the table side. Samuel Caley collection, Robert Grubb photograph.

MAD HATTER WITH WHITE RABBIT

The Mad Hatter leads with the White Rabbit attached to him. The Mad Hatter's clothes are yellow and green while the White Rabbit's clothes are yellow, red and blue.

Upper left: Mad Hatter With White Rabbit; upper right: Pebbles on Dino; lower left: The Big Bad Wolf with One Pig; lower right: Mickey and Donald on Alligator. Courtesy NEW ENGLAND AUCTION GALLERY, W. Peabody, Massachusetts.

MICKEY MOUSE AND LAWN MOWER

Mickey, dressed in red pants and a red cap, pushes a red lawn mower.

MICKEY AND DONALD RIDING ALLIGATOR

Mickey, dressed in yellow and red, and Donald, in his blue sailor suit, ride an alligator.

MICKEY AND MINNIE CARRYING A BASKET

Mickey in a yellow shirt and red pants, and Minnie in a green dress, hold a basket, the contents of which could be fruit.

MICKEY AND PLUTO

Pluto, in a crouching position as if tracking a scent, is followed by a gun-toting Mickey in a brown hunting outfit with a yellow cap. Embossed on the left side is the Walt Disney Productions copyright. The bottom of Mickey's right foot is rubber-stamped "Made In Hong Kong." This toy was designed by the talented Marx artist, Roy Ahlgren. This toy was initially rejected as being too risqué. Check the drawing in the Roy Ahlgren section and decide for yourself.

This 3 1/4" high plastic Ramp Walker was sold with instructions printed on a paper label. Samuel Caley collection, Robert Grubb photograph.

MINNIE MOUSE PUSHING A BABY CARRIAGE

A red-skirted Minnie pushes a tan baby carriage.

PLUTO

Pluto is in a standing position with his head lowered.

THE THREE LITTLE PIGS AND THE BIG BAD WOLF

The Three Little Pigs are followed by the Big Bad Wolf, all the figures joined together. Oddly, this toy is not shown on the instruction and advertising sheet mentioned in the introduction to this section. This sheet, which shows so many of the Disney Ramp Walkers, shows a version of the Wolf followed by just one Pig, plus another Ramp Walker in which there are only two Pigs. The wolf and single pig version was made, so possibly the other was too.

Value (For Ramp Walkers): Excellent $50-$75 each.

ROBIN HOOD AND FRIENDS #6807HK: 1974. About 4" high. Wind-up.

In these Marx Hong Kong plastic toys, three animal characters from the 1973 Disney film of the same name "dance" - which translated means that they vibrate. As described in the 1974 *MARX TOYS CATALOG*, page 41, "Robin Hood and Friends, 4" plus wind-ups, do twirling jig." No price is indicated.

Robin Hood, an orange fox with a white snout, is dressed in a brown outfit with a jaunty yellow cap. He carries a bow and arrows. Little John, a rather chubby brown bear with a pink snout, is dressed in a blue outfit with a matching blue cap. Friar Tuck is a similar bear but lacks the pink snout, his face being all brown. He is dressed in brown monk's clothing with a rope belt.

Value: ? ?

ROCKYKINGS: 1960s. 2 1/4" high. Non-powered.

This hard to find Marx Hong Kong plastic toy consists of Dopey mounted on a weighted base resembling half of an egg, which causes the toy to rock. It is believed that Donald Duck, Mickey Mouse, Pluto and Goofy were also made in this series.

Dopey has a very pink face and is dressed in a green outfit with a black belt, brown cap and pointed grey shoes. On his back is a gold paper label with "Walt Disney Productions/Hong Kong." Dopey stands on a yellow base.

The original box reads "Walt Disney's Rockykings/By Marx/Made In The British Crown Colony Of Hong Kong." The box top has the Marx logo and the box bottom has the Magic Marxie figure. Each box side shows a different character believed to have been made in this series, and each stands on a differently colored base. Donald stands on a blue base, Mickey also stands on a blue base, Pluto sits on a purple base and Goofy is on a green base.

Value: Excellent with box $50-$60

The 2 1/4" high plastic Marx Hong Kong Rockyking has a weighted base which causes it to rock. Samuel Caley collection, Robert Grubb photograph.

ROLYKINS: 1960s. 1 1/4" to 1 1/2" high. Ball-bearing action.

Rolykins are small Marx Hong Kong plastic toys with bell-shaped bodies. Inserted in the bases of the toys are steel ball-bearings which allow the toys to roll. The toys were manufactured using at least eight different characters.

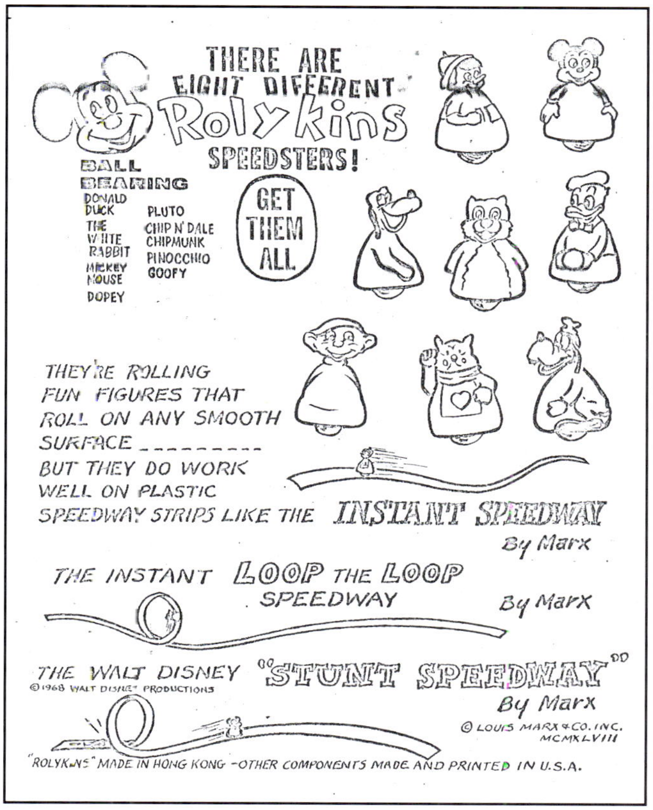

Rolykins original art by Marx artist Roy Ahlgren.

CHIP CHIPMUNK

The name of Dale, Chip's mate, was not used by itself, but oddly it was sometimes combined with Chip's name as in "Chip n' Dale" as if the whole name belonged to him. Chip is brown and white.

DONALD DUCK

Donald is dressed in a blue sailor suit and holds a red ball.

DOPEY

Dopey is dressed in green with a brown cap.

GOOFY

Goofy is dressed in a red outfit with a red hat.

MICKEY MOUSE

Mickey is dressed in red.

PINOCCHIO

Pinocchio is dressed in red, yellow and white.

PLUTO

Pluto has a red collar.

THE WHITE RABBIT

The White Rabbit from "Alice In Wonderland" is the most elaborate of the Rolykins, dressed in red and white with a red heart on his chest and one on his back, and a blue collar which is ruffled in back.

The Rolykins toys have gold paper labels with "Hong Kong" and the Walt Disney Productions copyright.

The Rolykins were packaged in separate boxes and also on blister cards. The boxes for all the characters are the same. The top flap reads "Walt Disney's Rolykins With Ball-Bearing Action" plus the name of the character. The same words are repeated on the bottom flap, along with the Marx logo and "Made In Hong Kong." The attractive multi-colored box has a different Disney character illustrated on each side. Shown are the heads of Mickey, Donald, Goofy and Pluto.

Value (For boxed Rolykins): $15-$25 each.

The blister cards on which the Rolykins were packaged came in at least two different designs, with different numbers of Rolykins attached to them. The card title is "Walt Disney's Rolykins With Ball-Bearing Action By Marx." Also on the card is "See if you can hit them & knock them over!" In the middle of the card is the Walt Disney Productions copyright. On the bottom of the card is the Marx name and logo and "Made In The British Crown Colony Of Hong Kong." Six different Rolykins (Mickey, Donald, Chip, Pluto, Goofy and Pinocchio) are attached to the card under plastic. The illustration makes the Rolykins appear to be riding on a speeding track. In the background are mountains and palm trees. Also illustrated on the card are Donald, Goofy and, in the distance, Mickey.

Value (For 6 pack blister Rolykins): $50-$75 each.

A similar blister card came with only two Rolykins attached to it and the two Rolykins could vary. This card reads "Walt Disney's Rolykins By Marx/The Figures That Run On A Ball-Bearing." A Stunt Speedway is mentioned, a toy that was made especially to go with the Rolykins (see next listing). Disney figures illustrated on the card are Mickey, Donald, The White Rabbit and Chip. On the lower part of the card, a looping track is illustrated, with one Rolykin rolling around the loop while the other rolls down the track.

Value (For 2 pack blister Rolykins): $15-$25

ROLYKIN STUNT SPEEDWAY: 1968. 12 foot long flexible track.

The track speedway mentioned above was made especially to go with the Rolykins. It consists of a 12 foot long flexible track, one anchor hook with rubber suction cup, and one or two loop clips 3 1/2" long each, all made of yellow plastic. Also included is one boxed Rolykin - the character could vary. The toy comes with an instruction sheet showing how to perform different types of tricks with various kinds of loops made out of the flexible track.

The original box reads "Walt Disney's Instant Stunt Speedway/Featuring The Thrilling Loop The Loop Action Starring One Of The Rolykins/The Little Figures That Run On A Ball-Bearing." The box is illustrated with various Rolykins rolling around loops and down the track. The back of the box features yet more tricks that can be done in addition to those mentioned in the instruction sheet.

Also on the back of the box is the statement that there are eight different Rolykins and it gives their names, including the combination designation for Chip, namely, Chip n' Dale. (The eight Rolykins are described in the listing immediately preceding this one.) A rather interesting note at the bottom of the box is "Rolykins Made In Hong Kong. Other Components [probably the ball-bearings and labels] Made And Printed In U.S.A."

Value: ? ?

Marx Hong Kong plastic Rolykins measure 1 1/4" to 1 1/2" high. They roll along on steel ball-bearings. Also shown is the Rolykin Stunt Speedway, 12 feet of flexible plastic track which was sold along with one Rolykin. Many tricks can be performed with the Rolykins and the Speedway. Samuel Caley collection, Robert Grubb photograph.

ROLY POLYS: 1960s. 5" high each. Non-Powered.

This rather odd-looking Marx plastic toy consists of a character's head attached to a ball-shaped body, with the character's hands resting on its stomach. The action of the toy is to roll back and forth when pushed. The toy is of the type a very young child might have played with.

The toy was made in different versions with either Mickey Mouse, Goofy or Pluto. A similar Roly Poly has been seen of the same size, but with a less rounded shape and bigger hands. Whether or not this similar toy was made by Marx is not known.

Value: Excellent with box $15-$25

SKY AKNE INDIAN SKATER: 1960s. 6 1/2" high. Wind-up.

Sky Akne, manufactured by Linemar, represents an Indian character from the 1961 Disney film "Babes In Toyland." This rare toy skates by using his right foot to push himself forward. The toy is made of tin with cloth clothing.

One of the most notable features of Sky Akne is his rather large nose, along with his fierce expression. The Indian carries a bow and arrow in his right hand. His clothing is tan with orange trimmed boots and a cream, orange and black headdress. Similar skating toys made by Linemar are Mickey Mouse Roller Skater and Roller Skater Popeye. (For descriptions, see Mickey Mouse toys and Popeye toys, respectively.)

The original box reads "Sky Akne Indian Skater From Walt Disney's Babes In Toyland." The scene illustrated is that of Sky Akne, roller skates on his feet, bow and arrow in hand, dancing around a fire along with other Indians. In the background are several wigwams.

A photograph of the toy and box with no description can be seen in an ad called "Happy Days" by the well-known dealers and collectors, Stephanie and Paul Sadagursky, Centerpark, NY, in the *COLLECTORS' SHOWCASE MAGAZINE,* July 1990, p.30.[9]

Value: With box $550-$600

SNOWDOMES: 1961. 3 1/4" - 4" high each. Non-powered.

Snowdomes consist of liquid-filled, translucent plastic globes containing various figures and scenes. When the Snowdomes are shaken, simulated snow falls inside. Marx Snowdomes are not as well known as some other Marx creations, but the company manufactured at least five different Snowdomes in both character and non-character versions. Each has a name sign at the base. Marx Snowdomes are not common and the character versions, in particular, are highly sought by collectors.

ALICE IN WONDERLAND

Alice, in a blue dress with a white apron, sits on the grass, her left arm supporting her. The White Rabbit, in red, white and blue with a red heart on his chest and his left arm raised, stands near her. In the background is a large tree under a rather dark blue sky.

BABES IN TOYLAND SOLDIER

The soldier, in a red and white uniform, with yellow trim and black boots, marches with a rifle under his arm. To his right is a smiling creature not easily identified. Nearby are a drum and a cannon. The background consists of a wall with two mounted flags flying, the sky and a large cloud in front of which is a balloon.

BRER FOX

Brer Fox, in a yellow shirt, red pants and a black vest, dances while Brer Rabbit, in a dark pink shirt and blue pants, watches. In the background are trees, billowing clouds in the sky, plus a log and flowers.

DONALD DUCK

Donald Duck sits in a car. In the background is a castle and mountains.

ONE HUNDRED AND ONE DALMATIANS

This scene consists of Pongo and Perdita, the male and female Dalmatian parents from the film, and just one of their puppies. Nearby is a house with trees in the background, and the sky.

PETER PAN

Peter Pan, dagger in hand, glares at Captain Hook and a sailor. Peter Pan is dressed in his customary green outfit with a green cap. Captain Hook wears a red suit with white trim and black boots. The sailor is dressed in a red and white shirt, red cap and blue pants. The three figures stand on a rocky shore while in the background can be seen Captain Hook's ship, craggy rocks sticking out of the sea, and the sky.

Most of the Snowdomes described above are shown in the excellent book *SNOWDOMES* by Nancy McMichael.[10]

Value: Excellent for all $100-$200

SNOW WHITE: See Tea Sets below.

SNOW WHITE TEA SET: 1950s-1960s. Size ? China.

This Marx uncommon child's play tea set probably came with six place settings. The set consists of large plates with various scenes from the Disney "Snow White" film, cups and saucers, a creamer and even a sugar bowl decorated with pictures of the Seven Dwarfs, and a teapot showing Snow White herself.

Except for the small plastic Snow White Disneykin figures (covered in a later volume) and the 1938 Dopey Walker (see *GREENBERG'S GUIDE TO MARX TOYS,* Volume I, pps.77-78) Marx items from the 1937 "Snow White," Disney's first animated feature film, are rare.

Value: $175-$250

WALT DISNEY 23 PIECE CHINA TEA SET: c.1958. Size ? China.

This Marx Japan child's play tea set has six place settings. The china pieces are white, each one decorated with one of nine scenes featuring various Disney characters. The set includes large dishes, cups, saucers, creamer, sugar bowl and teapot.

As described in the April 1990 *SMITH HOUSE TOYS CATALOG,* Eliot, ME, the lid of the original box reads "Walt Disney 23 (sic) Piece China Tea Set Plus 7 (sic) Punch-Out Stand-Up Figures." The figures are Mickey, Jiminy Cricket, Pluto, Minnie, Bambi, Lady The Dog and Thumper The Rabbit. Above the names of the figures, at the top edge of the lid, are die-cut, non-punch-out figures of Goofy, Tramp The Dog and Donald Duck.

Value: Excellent in box $600-$700

THUMPER THE RABBIT: 1950s-1960s. 3" long. Friction.

This tin toy is the Linemar version of Thumper, the cute little rabbit, who was Bambi's friend in the 1942 Disney animated feature film "Bambi." Thumper is grey with a yellow chest and white muzzle. He sits with his front legs stretched out in front of him.

Flower The Skunk, also from the film "Bambi," and Figaro The Cat, from Disney's animated film "Pinocchio," are similar toys, as is Pluto.

Value: Excellent $100-$110

A group of 1950s Linemar friction toys consisting of Figaro the Cat, Flower the Skunk, Pluto the Dog, and Thumper the Rabbit. Courtesy SMITH HOUSE, Eliot, Maine.

ZORRO ACTION SET: c.1957. Size ? Non-powered.

Zorro, a television series presented by Walt Disney, made his debut in 1957. In addition to the toy described here, Marx also manufactured a plastic Zorro figure on a horse (see after Zorro Action Set) and a Walt Disney's Zorro Playset.

The Zorro Action Set includes a hat, mask, whip, lariat, pistol, two fencing foils and a sheath knife.

The toy comes in a display box reading "Zorro Action Set/By Marx." Also shown is the Walt Disney Productions copyright. There is a listing of the various items included in the set. The dashing, masked Zorro figure is illustrated, cape swirling and sword drawn in his right hand.

Value: $25-$75

ZORRO ON HIS STALLION: 1958. 11" high.

The Zorro figure was shown in the *SEARS 1958 CHRISTMAS CATALOG* (page unknown), described as "Reproduced in Black Plastic with High Detail...Flexible Vinyl Accessories Are Detachable." The toy sold for the odd sum of $2.57. The ad does not mention what the accessories were, but they were probably Zorro's mask, hat and whip. The fact that the accessories are detachable accounts for their frequent loss; but even without the accessories, the beautifully modeled figure is quite handsome. Zorro is seated on his horse Midnight, whip in hand, with his cloak flaring out behind him.

Value: $250-$275

Zorro On His Stallion Midnight. Courtesy SMITH HOUSE, Eliot, Maine.

FOOTNOTES - DISNEY TOYS CHAPTER

1. *PLAYTHINGS,* April 1963, p.115.
2. *PLAYTHINGS,* June 1962, p.126.
3. Stephanie and Paul Sadagursky, "Happy Days" *ANTIQUE TOY WORLD,* April 1984, Vol. 14, No. 4, p.2.
4. Fred and Marilyn Fintel, *Yesterday's Toys With Today's Prices* (Lombard, Illinois: Wallace-Homestead Co., 1985), p.79.
5. Craig Ullery, "The Magic Of Mickey Mouse" *ANTIQUE TOY WORLD,* August 1986, Vol. 16, No. 8, p.75.
6. *PLAYTHINGS,* April 1961, p.117.
7. *PLAYTHINGS,* April 1962, p.131.
8. *PLAYTHINGS,* April 1954, p.126.
9. Stephanie and Paul Sadagursky, "Happy Days" *COLLECTORS' SHOWCASE,* July 1990, p.30.
10. Nancy McMichael, *Snowdomes* (New York: Abbeville Press, 1990), p.71.

Index

Other Books From Schiffer Publishing:

Collecting the Space Race Stuart Schneider. This book explores the artifacts of the dreams of the space age, beginning with the start of the 20th century. This book will inspire and answer some of the questions about the items that have been saved or collected as it exposes a broad field of things to collect. Over 550 color photographs. Price guide.
ISBN: 0-88740-535-5 $34.95

PEZ® Collectibles Richard Geary. After PEZ candy and dispensers were brought to the United States in 1952, over 250 different types of character heads were made. Most of the variations of PEZ dispensers are shown in 281 color photographs, as well as the premiums and store displays that helped sell them. The captions identify each item and give the current estimated values.
ISBN: 0-88740-693-9 $19.95

Hot Wheels: A Collector's Guide Bob Parker. This book provides extensive information about all the die cast vehicles that Mattel, Inc. made in this popular line from 1968 to 1990. Over 100 color pictures display nearly all the Hot Wheels made up to 1986, and a complete variation guide and checklist covers all the cars through 1990. Price guide.
ISBN: 0-88740-488-X $16.95

Baby Boomer Toys and Collectibles Carol Turpen. Beautiful color photos of over 780 toys and a helpful price guide cover favorites from robots, space toys to race cars, Hanna-Barbera toys, and Beatles collectibles. Adult children of the 1950s and 60s baby boom will feel they have rediscovered old friends and recaptured wonderful memories. Price guide.
ISBN: 0-88740-495-2 $29.95

A Trekker's Guide to Collectibles with Prices Jeffrey B. Snyder. The mountain of collectibles Star Trek has left in its wake are presented here in text, lists of items and over 490 color photographs with captions chronicling the field. Brief synopses of the series and films put the collectibles in perspective. Price Guide for over 3700 items. 465 color photographs.
ISBN: 0-88740-965-2 $29.95

Printed in China